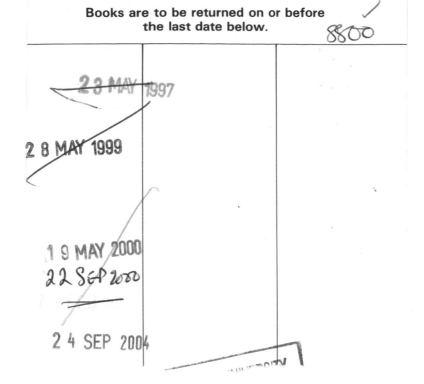

OXFORD STATISTICAL SCIENCE SERIES

SERIES EDITORS

J. B. COPAS   A. P. DAWID

G. K. EAGLESON   D. A. PIERCE

OXFORD STATISTICAL SCIENCE SERIES

Measurement, Regression, and Calibration

PHILIP J. BROWN

Department of Statistics and
Computational Mathematics
University of Liverpool

CLARENDON PRESS · OXFORD
1993

Oxford University Press, Walton Street, Oxford OX2 6DP

Oxford   New York   Toronto
Delhi   Bombay   Calcutta   Madras   Karachi
Kuala Lumpur   Singapore   Hong Kong   Tokyo
Nairobi   Dar es Salaam   Cape Town
Melbourne   Auckland   Madrid
and associated companies in
Berlin   Ibadan

Oxford is a trade mark of Oxford University Press

Published in the United States
by Oxford University Press Inc., New York

A catalogue record for this book is available from the British Library

Library of Congress Cataloging in Publication Data

ISBN 0 19 852245 2

Typeset by the author using LaTeX
Printed and bound in Great Britain by
Biddles Ltd, Guildford and King's Lynn

Preface

This book has been designed primarily as a research monograph for a range of regression problems in which one set of variables is predicted from another. It presupposes a familiarity with statistical theory as might be typically achieved in undergraduate courses. It is hoped that it will be of use in graduate statistical courses as well as a reference text for statisticians and scientists with some experience of statistical techniques. It is likely to be of particular interest to research scientists in industry and medicine.

The book starts with a range of examples and develops techniques progressively, starting with standard least squares prediction of a single variable from another and moving onto shrinkage techniques for multiple variables. Chapters 6 and 7 refer mostly to methods that have been specifically developed for spectroscopy. The other chapters are quite general in their applicability. Likelihood and Bayesian inference features strongly, the latter allowing flexible analysis of a wide range of multivariate regression problems. The last chapter presents some Bayesian approaches to pattern recognition.

For teaching purposes instructors may find particular chapters sufficiently self contained to recommend in isolation as reference or reading material. For example Chapter 4 gives an in depth development of a range of shrinkage techniques, including partial least squares regression, ridge regression and principal components regression, together with discussion of the recently proposed continuum regression. Chapter 8 on pattern recognition may also be of use by itself in courses on multivariate analysis and Bayesian statistics.

I began writing the book whilst on study leave at the Australian National University during the Australian summer of 1991/92. I am grateful to Liverpool University for allowing me to shelve my teaching and administrative duties for 6 months, to Peter Hall for the invitation to spend time at ANU, and to the Royal Society for travel funds. The manuscript has since been through some restructuring, modifications and additions. I am particularly grateful to Rolf Sundberg for his detailed and constructive comments. A number of others commented helpfully on sections of the book: I thank Philip Dawid, Timo Mäkeläinen, Mervyn Stone, Tom Fearn, and Anthony Atkinson. Ann Gould and statisticians at Shell Research, Sittingbourne, provided the invaluable sugars data, sponsorship and support for earlier work, and commented in detail then. I also thank SERC and the Complex Stochastic System initiative for funding two projects which supported Michael Denham and purchased SUN workstations and SPLUS software used throughout the book. I am also most grateful to Mike Denham for his help and advice on statistical computing matters.

The book leans heavily on earlier work, some of it done collaboratively. I am grateful to Sam Oman, Cliff Spiegelman and Jim Zidek in addition to Rolf Sundberg, Timo Mäkeläinen, and Mike Denham already mentioned.

I thank Chris West and Deborah Ashby for providing two medical datasets, and to Unilever Research plc for the detergent data. Mike Ashby provided advice on thermodynamics and the Forbes data.

P. J. Brown

Liverpool University
June 1993

Contents

For Jeremy and Emmeline
and to Anne

1
Introduction

1.1 Relating variables

This book is concerned with relating sets of variables. It mostly concentrates on regression models relating two sets of variables. The variables are measurements, or transformations of measurements, taken on an experimental unit. It is desired to predict one set of variables from the other set. Multiple, multivariate, linear, and non-linear regression models are analysed from a perspective of prediction.

In the most general setting there are $q \geq 1$ response variables and $p \geq 1$ explanatory variables. Here the terminology 'response' and 'explanatory' surmise a degree of causality. Such causality is unambiguous if in the training or calibration data the explanatory variables are strictly controlled in a statistically designed experiment. For instance, in a later example, the levels of three sugars, sucrose, glucose, and fructose, in an aqueous solution may be accurately controlled by formulation, taking a prescribed amount of each ingredient. If, however, both sets of variables are random such a direction of causality may be missing. In future, one set of variables, perhaps comprising traditional accurate measurements, is to be predicted from the other set of variables. This other predictor set may be much quicker and less expensive to obtain than the predictand set. Thus there is a natural asymmetry even when both sets of variables are random. Inferential problems are solved if it is possible to obtain the conditional distribution of the predictand given the predictor variables. In a controlled experiment, if the response is the predictand then inference is straightforward, as the training data provides observations on the conditional distribution of predictand given predictor. If, however, as is often the case the predictand is the explanatory variable set then the required conditional distribution is not directly available. It is then necessary to infer the explanatory set that gave rise to the observed response when only random departures from the conditional mean of the response given the explanatory set are measured by the training data. This will be called *controlled* calibration in contrast to the easier *random* calibration.

Both types of prediction will be considered throughout, but because of its inferential niceties and practical importance there is an extensive emphasis on controlled calibration. The book is also straightforwardly about regression in its normal theory form. Less straightforwardly it develops a range of modern techniques for coping with collinear data and more particularly regression with many variables. As such it has wide applicability in science, medicine, and industry.

1.2   Many variables

The predictor set of variables may comprise many variables chosen to relate linearly with the predictand. Even with careful prior consideration it may be hard to limit the number of variables and there are examples in this book where the number of variables is several orders of magnitude more than the number of observations in the training data. Many parameters of the full regression model will inevitably be inestimable, but it is still possible to accurately predict the predictand. Regression estimators which accommodate collinearity range from forms of principal components regression to adaptive ridge regression, partial least-squares regression, and continuum regression. The most flexible techniques arise out of a fully Bayesian analysis and incorporate realistic prior assumptions which supplement the data inadequacies.

1.3   Contents

Chapters 2 and 3 develop standard models for simple and multiple linear regression. Chapter 4 describes a range of regularization techniques for use in multiple regression, adaptive ridge regression, principal components regression, partial least-squares regression, and continuum regression. Theoretical and computational aspects are covered. In presenting these statistical approaches prediction is emphasized. Consideration is given later to the fundamental stage of model criticism. Hypothesis testing enters formally here and is somewhat de-emphasized in this book, except when discussing variable selection in Chapter 3; rather graphical exposure of departures from assumptions is emphasized.

   One common inferential tool used throughout is the likelihood function, and its variant after removal of nuisance parameters, the profile or maximum relative likelihood. This is used for generalized likelihood ratio confidence intervals and hypothesis testing from a sampling perspective. As a function, given the data, the profile likelihood is close to a Bayesian posterior distribution following diffuse prior information. Bayesian analysis with both diffuse and informative prior information is developed alongside more classical approaches in Chapter 5. Here a wide range of multivariate estimators, procedures, and diagnostics are developed and properties established. Chapter 6 considers methods which recognize the curve-like nature of data from some spectroscopic measuring instruments, where instead of a few responses one has a set of contiguous responses, a subsample of a continuous curve. The methodology in this chapter is exclusively Bayesian. It is subsequently used as an introduction to spatial models in environmental monitoring: a Bayesian alternative to kriging. Chapter 7 treats non-linear regression and calibration problems, and diagnostics for linearity and methods for response selection which avoid non-linearities. Chapter 8 considers the topic of pattern recognition or discrimination, from an entirely Bayesian perspective, again allowing for when there are fewer observations than variables. Pattern recognition relates continuous Y to discrete, nominal x, the identifiers of the populations, and clearly falls into the same paradigm as the rest of the book in wishing to predict x from Y.

There are appendices on multivariate distribution theory, Bayesian and conditional inference, Stein estimation and dominance of adaptive ridge estimators, matrix results, and an algorithm for partial least-squares regression.

1.4 Some datasets

Six typical datasets which are analysed later are introduced here. A number of other datasets are described more briefly to give a flavour of the wide range of applications. Some of these are also used illustratively at points in the text.

Example 1.1 Forbes' data

In the middle of the nineteenth century a Scottish physicist, J.D. Forbes, wanted to be able to estimate altitude above sea level from measurement of the boiling point of water. He knew that altitude could be determined from atmospheric pressure, measured with a barometer, with higher pressures corresponding to lower altitudes. In the experiment described here he collected both pressure and boiling point readings in seventeen locations in the Alps and Scotland. He was interested in the intermediate step of predicting pressure from boiling point. The data in Table 1.1 are from Forbes (1857), and are reported and analysed in Weisberg (1985); see also Atkinson (1985). The problem is modelled differently by these authors. They relate the logarithm of pressure to boiling point; we utilize the Clausius–Clapeyron's formula of classical thermodynamics, see Clausius (1850), to relate the logarithm of pressure to the *reciprocal* of boiling point (Kelvin). A plot of the data is given in Fig. 1.1, where it is clear that linearity on the chosen scales is not in question. As in the modelling of Weisberg and Atkinson, observation 12 appears to be anomalous, and was also dismissed by Forbes.

Whether the boiling point or pressure (or functions of either) ought to be treated as the explanatory x-variable is a debatable issue. You might, for example, argue that pressure is causative of boiling point from the physical point of view and therefore insist on turning around Fig. 1.1. However,

Table 1.1.   Forbes' data, giving boiling point (BP) (Fahrenheit) and barometric pressure (inches of mercury) for 17 locations in the Alps and Scotland

Case	BP(°F)	Pressure	Case	BP(°F)	Pressure
1	194.5	20.79	10	201.3	24.10
2	194.3	20.79	11	203.6	25.14
3	197.9	22.40	12	204.6	26.57
4	198.4	22.67	13	209.5	28.49
5	199.4	23.15	14	208.6	27.76
6	199.9	23.35	15	210.7	29.04
7	200.9	23.89	16	211.9	29.88
8	201.1	23.99	17	212.2	30.06
9	201.4	24.02			

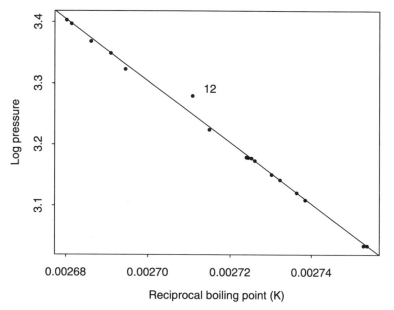

Fig. 1.1.   Forbes' data, log pressure versus reciprocal boiling point (Kelvin), with least-squares line

we will see that there are counter-considerations relating to whether the explanatory variable is subject to appreciable error.

Example 1.2 Penicillin dilution

Six concentrations of pure penicillin differing in twofold steps from 1 unit to 32 units per ml were set up on a plate (see Davies and Goldsmith (1984),

Table 1.2. Penicillin concentrations (in units per ml) and inhibition circle diameters (in mm)

Conc. penicillin	1	2	4	8	16	32
Circle diameter	15.87	17.78	19.52	21.35	23.13	24.77

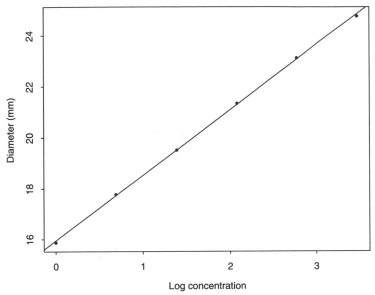

Fig. 1.2. Penicillin data, circle diameter (mm) versus log concentration (units per ml), with least-squares line

Example 7.3 Section 7.6). Table 1.2 gives circle diameters of the zones of inhibition in millimetres for each concentration. In this example the explanatory x-variable (that is, concentration of penicillin) has been set without appreciable error at fixed values. However, inaccuracies arise in the determination of the circle diameter. Figure 1.2 gives a graph of diameter against the logarithm of concentration, which is quite linear on this scale. In future you may wish to use a fitted relationship between diameter and log concentration to predict concentration from circle diameter. This may be needed even though concentration has been strictly controlled in the calibration and the temptation is to regress log concentration on diameter.

Example 1.3 Jug and syringe

This example from urology considers the true, jug, and syringe volumes for each of 24 haphazardly chosen (known) volumes of water which cover five 50 ml ranges between 10 ml and 260 ml (Haylen *et al.*, 1987). The

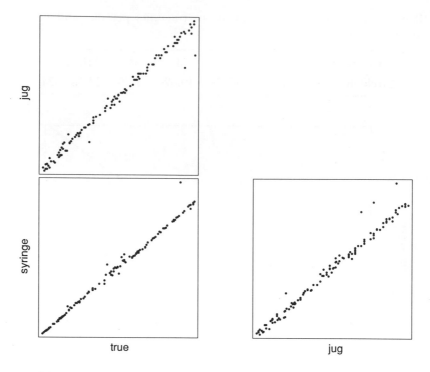

Fig. 1.3.   True, jug, and syringe volumes (ml) plotted in pairs

purpose of the comparison is to assess the accuracy of these measuring devices commonly used to measure the volume of fluid remaining in the bladder immediately following micturition, using catheterization. A 500 ml polypropylene jug and a 50 ml syringe are employed to independently estimate the true volume as measured by a calibrated cylinder (and checked by two observers). Figure 1.3 gives pairs of plots of the true volume, jug measurements, and syringe measurements.

Example 1.4 Bladder volume by ultrasound

A series of 23 women patients attending a urodynamic clinic were recruited for the study. After successful voiding of the bladder, sterile water was introduced in additions of 10, 15, and then 25 ml increments up to a final cumulative total of 175 ml. At each volume a measure of height (H) in mm and depth (D) in mm of largest ultrasound bladder images were taken. The product H × D was taken as a measure of liquid volume, albeit of areal dimension rather than a proper measure of volume. The true volumes together with the H and D product are given in Table 1.3 for the sequence of eight volumes (some incomplete) on the 23 women. For further description of the experimental protocol see Haylen *et al.* (1989). The product of

Table 1.3. Height–depth product of ultrasound bladder image for sequence of induced volumes, 10 ml, 25–175 ml in 25 ml steps with 23 women (NA = not available)

Volume induced (ml)							
10	25	50	75	100	125	150	175
13.2	27.4	41.6	60.1	73.9	85.5	95.6	107.4
11.1	27.5	58.1	78.8	91.5	98.3	111.4	121.0
10.3	15.0	34.2	49.4	71.3	81.3	94.0	104.3
NA	10.0	28.8	46.4	54.8	69.4	73.9	NA
4.8	18.6	29.9	39.4	NA	NA	NA	NA
7.7	12.6	31.4	45.3	48.0	66.6	NA	NA
NA	24.0	46.9	50.4	67.8	81.0	91.2	99.8
5.9	28.4	44.4	70.7	89.4	105.8	113.5	127.3
1.9	12.5	26.8	54.4	63.1	83.5	114.5	124.0
6.5	16.7	30.6	41.8	49.6	60.8	80.1	87.1
19.8	29.6	51.7	72.2	81.9	95.1	115.4	NA
14.6	27.1	49.8	67.5	79.1	95.1	109.8	NA
NA	14.0	19.1	39.2	48.7	67.0	72.7	NA
NA	18.7	35.8	49.6	65.6	85.3	90.4	NA
9.7	20.3	38.9	65.1	65.1	86.9	98.6	107.2
17.2	35.8	41.4	69.7	81.9	96.6	115.0	117.0
10.6	23.6	49.9	67.7	87.7	89.3	108.0	114.8
19.3	37.4	58.6	73.7	79.4	102.6	110.9	122.4
8.5	31.3	54.8	78.3	93.0	NA	NA	NA
6.9	23.7	44.0	65.7	80.3	93.6	99.2	112.2
8.1	22.0	39.1	44.7	68.9	93.3	102.4	104.7
14.8	34.3	58.5	72.1	90.9	105.0	117.5	124.2
13.7	28.5	41.5	59.8	77.5	92.9	99.4	113.0

H and D is plotted against true volume in Fig. 1.4.

Examples 1.5 and 1.6 Composition by infrared spectrum

The infrared region can be sub-divided into the near-, middle (or mid-), and far-infrared, and altogether covers that part of the electromagnetic spectrum between about 0.78 μm (12,800 cm^{-1}) and 100 μm (10 cm^{-1}). The mid-infrared region 2.5 to 50 μm has been successfully used by chemists for many years for the structural analysis of organic and inorganic compounds by assigning absorption bands to fundamental vibrations of the investigated molecule. The near-infrared spectra absorptions belong to overtone and combination vibrations but allow greater path lengths and much easier preparation of samples. Light fibre probes transmit absorptions to be measured at a considerable distance from the spectrometer. Linear statistical

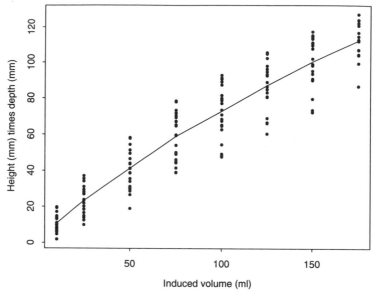

Fig. 1.4. Bladder data, height times depth versus volume, with lines joining averages at each volume

models relating absorbance to composition are underpinned by Beer's law (see Cross and Jones (1969)) which stipulates a linear relationship between the concentration of a substance dispersed in a non-absorbing medium and the amount of light absorbed by it. This is only an experimental law and departures from linearity occur for a variety of reasons, including interaction effects between components and with the medium. In the examples the composition of the sample preparations is accurately determined by formulation. In future you may wish to predict the composition from the spectrum of a presented solution.

Two examples of the use of infrared (IR) spectroscopy are provided, one mid-IR and one near-IR. Traditionally spectroscopists index mid-IR spectra by frequency but near-IR by wavelength. The mid-IR data (Example 1.5) concern the composition of a liquid detergent. Twelve sample preparations were chosen to have a range of compositions with four compounds in aqueous solution as given in Table 1.4. For each of the twelve samples the mid-IR spectrum was recorded as the absorbances at 1168 equally spaced frequencies (channels) in the range 3100 to 750 cm^{-1}. The corresponding twelve spectra are presented graphically in Fig. 1.5, where the absorbances emerge as a continuous curve to the resolution of the plotting device.

The near-IR data (Example 1.6) concern the composition of three sugars: sucrose, glucose, and fructose in aqueous solution. Here there is a full 5^3 design of compositional data, each sugar being at levels 6, 10, 12, 14,

Table 1.4. Twelve sample compositions of detergent, four compounds in water (percentage by weight)

Comp. 1	Comp. 2	Comp. 3	Comp. 4	Water
20.00	6.99	3.02	6.35	63.64
14.92	13.10	5.29	6.03	60.67
19.95	9.91	5.39	4.00	60.75
24.65	13.09	3.01	4.00	55.25
24.91	7.08	5.24	8.36	54.40
14.87	6.89	7.12	4.09	67.03
25.20	10.07	6.95	6.07	51.71
15.12	10.08	3.46	8.19	63.14
19.92	13.22	9.38	7.96	49.52
21.39	7.91	6.28	6.16	58.25
17.65	12.12	6.43	7.17	56.63
16.69	9.49	4.07	4.70	65.07

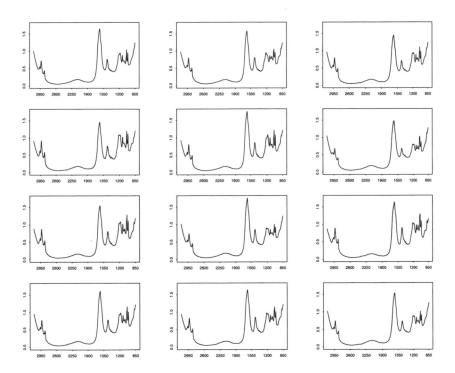

Fig. 1.5. Detergent data, absorbance spectra of 12 samples (by row) by frequency (cm^{-1})

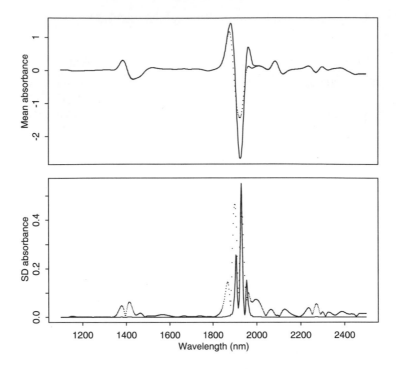

Fig. 1.6.   Sugar data, absorbance (2nd difference) spectra sample mean and standard deviation, for 125 observation training data (solid) and 21 observation validation data (dotted) by wavelength (nanometers)

Table 1.5.   Sugar validation data, percentage composition of three sugars in water

Sample	1	2	3	4	5	6	7	8	9	10	11	12
Sucrose	0	25	25	0	0	25	25	12	12	0	25	12
Glucose	0	0	0	25	25	25	25	12	12	12	12	0
Fructose	25	25	0	0	25	25	0	0	25	12	12	12

Sample	13	14	15	16	17	18	19	20	21
Sucrose	12	12	25	12	0	12	25	12	0
Glucose	25	0	12	25	12	0	12	25	12
Fructose	12	0	0	0	0	25	25	25	25

and 18 per cent by mass. For each of the 125 samples a second derivative spectrum of 700 absorbances at frequencies corresponding to wavelengths 1100 nanometers (nm) to 2498 nm in steps of 2 nm was obtained. A validation or prediction set was also available and had sugar concentrations at three levels: 0, 12, and 25 per cent by mass and was a subset of 21 of the full $3^3 = 27$ design, perversely largely outside the range of the calibration design (see Table 1.5). Figure 1.6 gives the sample mean and standard deviation as a function of wavelength for calibration and validation data, averaging over 125 and 21 observations, respectively. The spectral curves are evidently quite similar for all 146 observations, with the largest variation where the mean curve changes most, at around 1900 nm. This is where water dominates.

Other examples that are referred to are:

1. Measuring Marathon courses by means of bicycle counters as in the Los Angeles Olympics (see Smith and Corbett (1987)).
2. Estimating fetal age by ultrasound measurements (see Oman and Wax (1984)).
3. Measuring the age of a Kangaroo (see Wood *et al.* (1983)).

2
Simple linear regression

2.1 Estimation and model criticism

Suppose a single random variable Y is such that for a given value of another variable $X = x$, the conditional expectation is linear in x :

$$E(Y|X = x) = \alpha + \beta x. \tag{2.1}$$

Here Y might be some measure of lung function and x the cumulative exposure to asbestos dust for a building worker. We may note that x may not be the originally measured variable, it may rather be some function of it; the same applies to Y. What is required is that linearity of conditional expectation holds for the calculated variables, and that in obtaining a measurement pair (x_i, y_i) on each of n workers, uncorrelated errors ϵ_i, with constant variance, add to eqn (2.1) to give the random observation Y_i, so that

$$Y_i = E(Y|X_i = x_i) + \epsilon_i. \tag{2.2}$$

Implicitly ϵ_i must be uncorrelated with x_i. Also implicitly the selection of building workers has been made by some random mechanism. At most the mechanism could depend on exposure to asbestos and should certainly not depend on lung function.

This is the simple linear regression model with usual second-order error assumptions. For probabilistic inference (for example, confidence intervals) typically normality is assumed, although other distributional assumptions may arise. Point estimation of the unknown parameters $(\alpha, \beta, \sigma^2)$ (intercept, slope, error variance) requires only second-order assumptions. Estimates of (α, β) which minimize the residual sum of squares $\sum (y_i - \alpha - \beta x_i)^2$ are justified by the Gauss–Markov theorem. This states that least-squares estimates are Best (minimum variance), amongst Linear, Unbiased, Estimators (BLUE) under the above model.

The model, its second-order assumptions, and normality need to be examined critically on both prior and empirical grounds. The substantive context of the data is important, and, for example, Popper (1959) emphasizes that data need to be theory laden. Theory and data complement one another and are both indispensable in order to arrive at a valid model. The data itself can be made to confront aspects of theory with various data and residual plots.

In the Forbes data (Example 1.1) suppose we take Y to be log(pressure) and x be reciprocal of boiling water in Kelvins. Here it is enough to know that this relationship can be expected to be linear (from the Clausius–Clapeyron equation of thermodynamics). A graph of the 17 data points in Fig. 1.1 does indeed look linear. It has one likely outlier, and this outlier is the focus of attention for graphical diagnostics for this example by Weisberg (1985) and Atkinson (1985). Their model relating boiling point in degrees Fahrenheit to log of pressure does not make prior physical sense, however. Over the limited range of Forbes' data, the relationship appears linear on both scales, but the physical law is helpful for extrapolation and interpretation (the slope in Fig. 1.1 well approximates $-R$, where $R = 4860$ is the gas constant, the heat of evaporation of water around 373 K).

Model (2.2) is asymmetric in its treatment of x and Y, as is also the least-squares fitting procedure. A fundamental issue is whether it might be better to relate pressure to boiling point, or the other way around, irrespective of whether or not on reciprocal or logged scales. After all, one might argue that changes in pressure cause the boiling point of water to change. Whilst this view is quite tenable, there are two distinct reasons why one might still explain pressure by boiling point. Firstly, if the boiling points were accurately determined, whereas pressure determinations are subject to error, then one would wish to regress pressure on boiling point. However, reading Forbes' description of his measuring instruments it is not clear that his determination of boiling point is so much more accurate than his determination of pressure. One may therefore feel it necessary to fall back on the second reason. This suggests that either regression is appropriate since the pair, pressure and boiling point, are jointly randomly distributed. For future prediction of pressure from boiling point it is natural to then prefer the regression of pressure on boiling point. It is the essence of Forbes' data that he went to a variety of locations in the Alps and Scotland. He did not fix or control the pressures in advance, at least not directly, although he did in some informal way choose the locations and hence the altitudes. Cautiously, I suggest that his data may therefore be regarded as a sample of size 17 from a joint distribution of pressure and boiling point. It is random in the restricted sense of the protocols that went into his choice of fairly arbitrary places in Scotland and the Alps. If the conditional distribution of logged pressure given reciprocal boiling point is modelled and used for future prediction of pressure from temperature, then we will need to be conscious of the implicit ingredient: that future pressures ought to be from the same marginal distribution of pressure as in the experimental data. If on the other hand the new readings of boiling point were from the higher Himalyas then one would be more inclined to model reciprocal boiling point as Y with log pressure as x, and use

ESTIMATION AND MODEL CRITICISM

ESTIMATION AND MODEL CRITICISM

their fitted relationship. Then to predict pressure (and hence elevation) from boiling point one would invert this relationship. These issues are difficult and pointers can be offered at a more formal level later. Whether to regress Y on x or x on Y to be able to predict x in future has been a contentious issue over recent decades. We should also worry about applying the model outside the range of the calibration data, although the physical law underpinning it would make such extrapolations less risky.

The formulae and results for least-squares estimation are first summarized. Estimates of slope and intercept are

$$\hat{\beta} = S_{xy}/S_{xx}, \quad \hat{\alpha} = \bar{y} - \hat{\beta}\bar{x} \tag{2.3}$$

where $S_{xy} = \sum(x_i - \bar{x})(y_i - \bar{y}) = \sum(x_i - \bar{x})y_i$, $S_{xx} = \sum(x_i - \bar{x})^2$. Thus the least-squares line goes through the point $(\bar{x}, \bar{y})$ and has slope $\hat{\beta}$. Both $\hat{\alpha}$ and $\hat{\beta}$ for fixed $x_1, \ldots, x_n$ are linear in the random variables $Y_1, \ldots, Y_n$. They have the bivariate covariance matrix $\sigma^2(\mathcal{X}^T\mathcal{X})^{-1}$, where $\mathcal{X}$ is the $(n \times 2)$ model or carrier matrix:

$$\begin{pmatrix} 1 & x_1 \\ \vdots & \vdots \\ 1 & x_n \end{pmatrix}.$$

Common usage often refers to this as the *design* matrix but this term is perhaps best reserved for the variables that are observed or designed, such as the second column of $\mathcal{X}$. Here $x = (x_1, \ldots, x_n)$ is assumed not to be proportional to the unit vector, that is, the rank of $\mathcal{X}$ is 2 so that α and β are separately estimable. Thus $(\mathcal{X}^T\mathcal{X})^{-1}$ takes the form

$$1/(nS_{xx}) \begin{pmatrix} \sum x_i^2 & -\sum x_i \\ -\sum x_i & n \end{pmatrix},$$

and the variance of $\hat{\beta}$ is σ^2/S_{xx} and $\hat{\beta}$ is uncorrelated with $\hat{\alpha} + \hat{\beta}\bar{x}$ whose variance is σ^2/n. This implies that a parametrization of model (2.1) such that $\sum x_i = 0$ would result in independent estimators of $\hat{\alpha}, \hat{\beta}$ under normality. This just reflects the geometry that if two columns of a design matrix are orthogonal then corresponding estimators of parameters are independent under normality, and this is true in multiple and multivariate regression as well. The residual vector $e = (e_1, \ldots, e_n)$, where $e_i = y_i - \hat{y}_i$ with

$$\hat{y}_i = \hat{\alpha} + \hat{\beta}x_i,$$

is orthogonal to the columns of $\mathcal{X}$, so that in particular, $\sum e_i = 0$. More importantly the orthogonality of the residual and estimation spaces implies

Table 2.1. Analysis of variance for simple linear regression of Y on x

Source of variation	Sum of squares	Degrees of freedom	Mean square	F ratio
explained (by x)	$S_b = \sum(\hat{y}_i - \bar{y})^2$	1	$M_b = S_b/1$	$F =$
residual	$S_e = \sum(y_i - \hat{y}_i)^2$	$\nu = n - 2$	$M_e = S_e/\nu$	M_b/M_e
Total	$S_{yy} = \sum(y_i - \bar{y})^2$	$n - 1$		

that $\sum e_i^2$ is distributed independently of $\hat{\alpha}, \hat{\beta}$, as a chi-squared random variable, scaled by σ^2, on $(n-2)$ degrees of freedom. Thus $\hat{\sigma}^2 = \sum e_i^2/(n-2)$ is an unbiased estimator of σ^2 and the pivotal quantity,

$$\sqrt{S_{xx}}(\hat{\beta} - \beta)/\hat{\sigma}, \qquad (2.4)$$

has a Student-t distribution on $(n-2)$ degrees of freedom, and may be used for testing hypotheses or setting confidence intervals for the slope of the relationship of Y to x.

The sample correlation is defined as $r = S_{xy}/\sqrt{S_{xx}S_{yy}}$ and there are two regression lines whose slopes relate through r. We may note that the slope from regressing Y on x is $\hat{\beta} = r\sqrt{S_{xx}/S_{yy}}$. If x is regressed on Y but the fitted line is drawn on the same graph with this x-axis horizontal, the slope of the line which minimizes squared errors in the horizontal direction is $S_{yy}/S_{xy} = 1/\hat{\beta}_*$, with $\hat{\beta}_*$ the coefficient in the regression of x on Y. Thus the slope for Y on x is the product of r^2 and the slope of x on Y. Since $r^2 \leq 1$ the slope of Y on x is always less than or equal to that of x on Y. If Y and x are standardized then the slope of the regression of standardized Y on standardized x is r and has modulus less than 1; hence the term *regression* as used by Galton (1889) when studying heights of sons given their fathers' heights. When r^2 is near to 1 then the two regression lines are rather close.

The quantity r^2 is also both the proportion of Y-variation explained by x and the proportion of x-variation explained by Y. The proportion explained can be made to be the basis of an analysis of variance table as in Table 2.1. Here the F-test is on $(1, n-2)$ degrees of freedom and is used to test the null hypothesis $\beta = 0$. This test is exactly equivalent to the Student two-sided t-test from the pivotal quantity (2.4). The proportion of variation explained is a descriptive measure: it does not assess allowable uncertainty in the same way as the F- or t-test. For example, with only $n = 2$ observations the value of r^2 is 1 since a straight line may be exactly

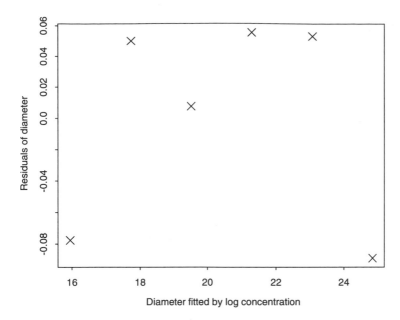

Fig. 2.1.   Penicillin data, residual versus fitted diameters for simple linear regression

fitted to two points in the xy plane, there being no degrees of freedom left over to assess error.

The normality and second-order error assumptions are usually examined by a variety of residual plots:

1. residuals versus fitted values;

2. residuals versus index order of observations;

3. residuals versus residuals of candidate further explanatory variable regressed on x;

4. residual normal quantile–quantile plot.

The first of these looks for changes in error variance with fitted value, or other features of the residuals related to fitted value. The second can be useful if experimental protocols change through time, producing drift in the residuals. The third is beneficial when the fit is inadequate and a second explanatory variable is suspected of influencing the response Y. It may be noted that if one formally regresses the residual from the fitted model on the residual of candidate explanatory variable, formed by regressing it on the explanatory variables already in the model, then an exactly appropriate test ensues. It corresponds to normal theory testing in multiple regression. It is sometimes called an *added variable test* and the graph *an added variable plot*.

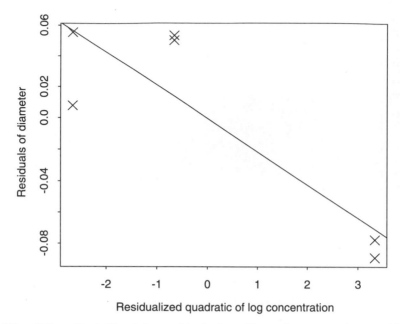

Fig. 2.2.   Penicillin data, residuals from linear fit regressed on residualized quadratic log concentration

The plots may also show up the occasional outlier, and here to avoid masking of features it is advisable to use leave-one-out or jackknifed residuals in the above plots. There may also be observations which have a particularly strong influence on the fitted regression coefficient, which may be assessed by jackknifed estimates of β. For a thorough discussion of these aspects see Cook and Weisberg (1982) and Atkinson (1985) and the pioneering work of Hampel *et al.* (1986). As illustration Fig. 2.1 plots residuals versus fitted diameters for the Penicillin dilution data of Example 1.2. Despite the high and statistically significant explanation of variation by the line with $r^2 = 0.9996$ (see earlier Fig. 1.2) there is a suggestion of slight concave dependence in the residuals. Indeed, a regression of the residuals from the linear fit on the quadratic of logged concentration, chosen so as to be orthogonal to the linear contrast and constant, is significant at the 5 per cent level. The relationship is plotted in Fig. 2.2 and constitutes an added variable plot, the added variable being a residualized quadratic.

Figure 2.3 gives an index plot of the residuals for the Forbes example 1.1, with the evident outlier of observation number 12. Figures 2.4 and 2.5 give normal quantile–quantile plots of the residuals from regressing jug and syringe volumes, respectively, on true volume for Example 1.3. The jug data show three outlying values. The syringe data suggest one outlying value and a distribution somewhat longer tailed than normal.

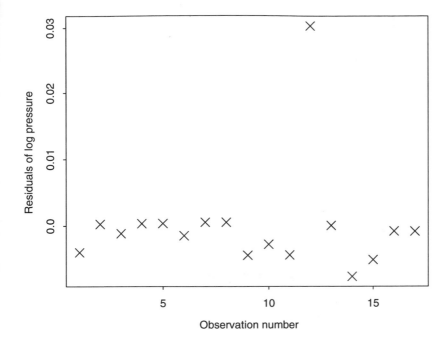

Fig. 2.3. Forbes' data, index plot: residuals from linear fit of log pressure on reciprocal boiling point (K), versus observation number

There are various viable strategies for dealing with non-normality and outliers. Outliers always deserve some attention when flagged, be it merely to look for possible causes by going back to the original experiment. Robust methods following Hampel *et al.* (1986) seek to minimize the impact of outliers whilst drawing attention to their presence. An outlier may be more interesting than the other data. Non-normality may be modelled or reduced by a suitable transformation (see Atkinson (1985)). However, one can often live with modest amounts of non-normality, at least when estimation entails sufficient averaging to allow the central limit theorem to take effect. The effect of having variance as a function of the mean may be a more important motivation for a transformation, as discussed in Section 2.7.

2.2 Prediction intervals for the response

Suppose one wishes to predict a future value of Y, denoted by Z, at a known fixed or given value x_0 of the explanatory variable. Assume that Z has the same model characteristics as Y, as given by eqns (2.1) and (2.2).

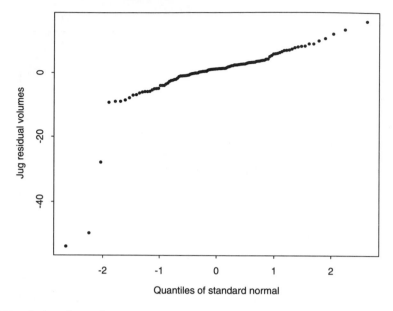

Fig. 2.4.   Jug volumes, normal quantile–quantile plot of residuals from linear regression on true volume

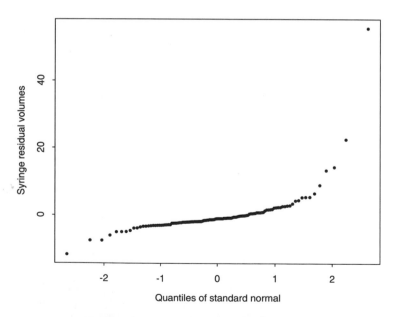

Fig. 2.5.   Syringe volumes, normal quantile–quantile plot of residuals from linear regression on true volume

The expectation of Z given $X = x_0$ is

$$E(Z|x_0) = \alpha + \beta x_0 \qquad (2.5)$$

and may be estimated by

$$\hat{Z} = \hat{\alpha} + \hat{\beta} x_0$$

with estimates from eqn (2.3), and variance

$$\text{Var}(\hat{\alpha}) + x_0^2 \text{Var}(\hat{\beta}) + 2x_0 \text{Cov}(\hat{\alpha}, \hat{\beta})$$

or

$$\sigma^2 (1, x_0)(\mathcal{X}^T \mathcal{X})^{-1}(1, x_0)^T.$$

The model for Z is

$$Z = E(Z \mid x_0) + \epsilon, \quad E(\epsilon) = 0, \quad \text{Var}(\epsilon) = \sigma^2. \qquad (2.6)$$

The mean-squared error of the predictor $\hat{Z}$ is

$$E(Z - \hat{Z})^2 = \sigma^2[1 + 1/n + (x_0 - \bar{x})^2/S_{xx}] = \sigma^2 c^2(x_0) \qquad (2.7)$$

and this defines the predictive scale factor $c(x_0)$, the square root of a quadratic in x_0. The first term in eqn (2.7) arises from the error variance of future Z and remains when, as $n \to \infty$, the second two terms vanish. This error variance for the future Z may in practice be larger than σ^2 since greater care will have been taken with the experimental conditions in the initial experimental data and it is wise to be alert to this possibility. It may entail collection of some (x, Y) in on-site or non-experimental conditions. Under normality the predictive distribution of Z for fixed error variance σ^2 is such that $Z - \hat{Z}$ is normal with zero mean and variance $\sigma^2 c^2(x_0)$. For σ^2 estimated by $\hat{\sigma}^2$, then

$$(Z \doteq \hat{Z})/[\hat{\sigma} c(x_0)] \qquad (2.8)$$

has a Student t-distribution on $(n - 2)$ degrees of freedom. The random variable in formula (2.8) is a predictive pivot for Z as a function of x_0, since it involves x_0, but has a fixed known distribution not involving x_0 or the parameters $(\alpha, \beta, \sigma^2)$. This sampling distribution is for repetitions from both the populations of Z and $\{Y_1, \ldots, Y_n\}$ for fixed x_0 and $\{x_1, \ldots, x_n\}$. The Student distribution does not entail any of these x-values so that the predictive distribution of Z is also unconditional. It will be seen later that this Student predictive distribution is exactly the same as a (vague prior) Bayesian predictive distribution for Z, but interpreted as conditional on all the data, x_0 and $\{(x_i, y_i), i = 1, \ldots, n\}$.

A sampling theoretic justification for focussing on formula (2.8) to set confidence limits on x for observed Z is in terms of a notional test of the equality of the parameters (α, β) in the distribution of Z and the distribution of $(Y_1, \ldots, Y_n)$ (see Cox and Hinkley (1974, p. 243)).

From formula (2.8) a $(1 - \gamma)100$ per cent prediction interval for Z is

$$\hat{Z} - t^*_{n-2}(\gamma)\hat{\sigma}c(x_0) \leq Z \leq \hat{Z} + t^*_{n-2}(\gamma)\hat{\sigma}c(x_0), \qquad (2.9)$$

where $t^*_{n-2}(\gamma)$ is the tabulated upper $(\gamma/2)100$ per cent point of the Student t-distribution on $(n-2)$ degrees of freedom.

Clearly such prediction intervals rest on the validity of the normal assumption. This reliance is more critical than in parameter estimation where data averaging occurs and the central limit theorem can be invoked. Observation number 12 of the Forbes data may be an outlier on even the most stringent tests, yet perhaps it is desired to predict the possibility of such occurrences, with a frequency of one in seventeen observations. If we delete observation 12 as most statisticians would be inclined to do, then we must admit that only mainstream future observations are to be predicted. Distributions with longer tails than the normal may be needed to truly reflect the data generation mechanism.

2.3   Natural and controlled calibration

Suppose a future value of the response $Z = z$ is observed but not the value of x that gave rise to this. This unknown value of x is hereafter denoted by the Greek letter ξ. There are two main cases to distinguish.

Firstly in some applications the training sample, $x_1, \ldots, x_n$, and ξ together form a random sample from a population. This constitutes knowledge of ξ prior to observing $Z = z$, and may with advantage be simply incorporated into our inference for ξ after observing $Z = z$. This is often called the *natural* or *random* calibration case. The best way to predict ξ is to reverse the roles of x and Y, that is, regress x on Y in the n observation training data and use the predictive interval given in eqn (2.9) with $x \to z$, $Z \to \xi$, and $(\hat{\alpha}, \hat{\beta}, \hat{\sigma}^2) \to (\hat{\alpha}_*, \hat{\beta}_*, \hat{\sigma}^2_*)$ least-squares estimates from the regression of x on Y. As already discussed the two estimated regression coefficients satisfy

$$\hat{\beta}\hat{\beta}_* = S^2_{xy}/(S_{xx}S_{yy}) = r^2$$

(The ratio of the regression slopes, drawn on the same xy-axes, is the squared correlation). The estimator of ξ is given as

$$\check{\xi} = \hat{\alpha}_* + \hat{\beta}_* Z,$$

or in centred x, y units (with Z adjusted conformably)

$$\check{\zeta} = (S_{xy}/S_{yy})Z. \tag{2.10}$$

In this context of random calibration, X is a random variable with (X, Y) jointly distributed. From this one is perfectly at liberty to construct the conditional distribution of X for given $Y = y$ in order to predict an unknown X (that is, ξ) in future. For distributions other than normal this is also the natural route.

If on the other hand $x_1, \ldots, x_n$ are fixed by design in the training experiment, and there is no guarantee that the unknown ξ can be assumed 'like' the controlled x-values, then one is restricted to inference from the conditional distribution of Y for given x. This is termed the *controlled* calibration case and gives rise to some inferential niceties, which the rest of this chapter will address. This is done both in the above case of no available information about ξ *a priori* and later on with the help of some random X linked to ξ. A Bayesian analysis linking ξ and the x-values is left to Chapter 5.

For controlled calibration the regression of Y on x is utilized. This is the favoured route, at least since Eisenhart (1939). He contended that one should minimize the errors in the direction they occur, namely the Y-direction. If ξ were known then the fitted value of Z would be $\hat{\alpha} + \hat{\beta}\xi$. The value $Z = z$ is observed and ξ estimated by solving the equation $z = \hat{\alpha} + \hat{\beta}\xi$ giving the estimator $\hat{\xi} = (Z - \hat{\alpha})/\hat{\beta}$, or in centred x, y units,

$$\hat{\xi} = (S_{xx}/S_{xy})Z. \tag{2.11}$$

The two estimators, $\hat{\xi}$ and $\check{\xi}$, are related by

$$\check{\xi} = r^2 \hat{\xi}. \tag{2.12}$$

They are presented graphically in Fig. 2.6 for the bladder data of Example 1.4 with $Z = 110$. The horizontal line drawn from $Z = 110$ cuts the two regression lines at two different values along the x-axis marked by a diamond and triangle for $\check{\xi}$ and $\hat{\xi}$ respectively. It is worth noting here that eqn (2.12) implies that $\check{\xi}$ is always closer to the mean of the calibration x than $\hat{\xi}$. The estimation is conditional on the assumption of a straight line relationship and is questionable for the bladder data, especially at the ends of the range of x-values. This is investigated further in Section 2.7.

The creation of a confidence interval is more problematic in the controlled case. If ξ were known the prediction interval for Z based on the distribution of formula (2.8) would be given by eqn (2.9) with ξ replacing x. We may construct a $(1 - \gamma)100$ per cent confidence 'interval' for ξ by taking those values of ξ such that

$$|(Z - \hat{\alpha} - \hat{\beta}\xi)/[\hat{\sigma}c(\xi)]| \leq t^*_{n-2}(\gamma). \tag{2.13}$$

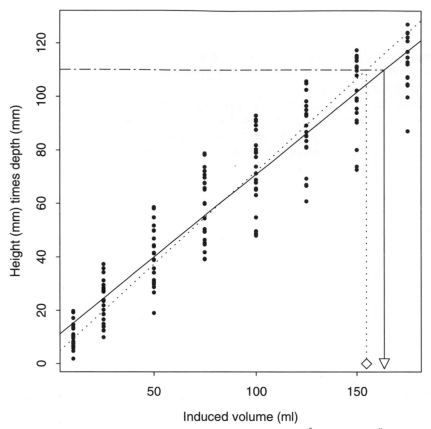

Fig. 2.6. Bladder data, two linear regressions and $\hat{\xi}$ (triangle), $\check{\xi}$ (diamond) for $Z = 110$

Apart from its natural derivation from the distribution of formula (2.8), a justification for this interval is that it corresponds to the inversion of a similar test of the hypothesis $\xi = \xi_0$ (Cox and Hinkley (1974), p. 268). It is closely related to Fieller's solution to the problem of inference about the ratio of two means. Here, however, it is later justified as a likelihood ratio confidence region. Squaring and rationalizing (2.13) requires the solution of a quadratic in ξ; taking $\bar{x} = 0$ and $t^*_{n-2}(\gamma) = t$ this gives

$$\xi^2(\hat{\beta}^2 - \hat{\sigma}^2 t^2/S_{xx}) - 2\xi\hat{\beta}(Z - \hat{\alpha}) + (Z - \hat{\alpha})^2 - \hat{\sigma}^2 t^2(1 + 1/n) \leq 0.$$

This is of the form $a\xi^2 + b\xi + c \leq 0$. Provided $a > 0$ the quadratic is as in Fig. 2.7 (i) or (ii), where the set of values for ξ in (i) is obtained from the end points, these being roots of the quadratic

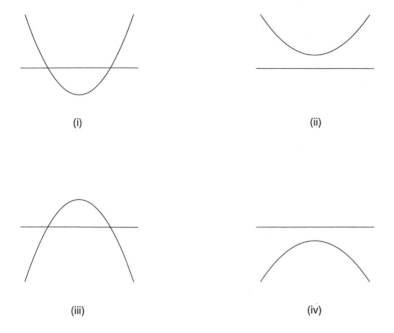

(i)

(ii)

(iii)

(iv)

Fig. 2.7. Confidence interval for ξ: four possible quadratic forms

$$[-b \pm \sqrt{b^2 - 4ac}]/(2a). \qquad (2.14)$$

In case (ii) the quadratic has no real roots and '$b^2 < 4ac$' implies that the set is empty. If $a < 0$ the parabola would flip over to (iii) or (iv). In case (iii) the region is every ξ value *apart* from the interval between the two real roots! In case (iv) the interval is the whole real line. Thus only case (i) provides a satisfactory confidence interval. We know for sure that the whole real line covers the true ξ with probability 1 , and the empty interval covers it with probability zero, yet the average coverage property encompassing situations (i), (ii), (iii), and (iv) is correctly $1 - \gamma$. In practice pathologies like (ii)–(iv) rarely occur. Suppose the slope is statistically different from zero so that

$$|\hat{\beta}\sqrt{S_{xx}}/\hat{\sigma}| > t^*_{n-2}(\gamma). \qquad (2.15)$$

Any self-respecting calibrator will design a calibration experiment such that his or her instrument is expected to have a statistically significant slope in relating measurement to truth. Thus provided expectations are met $a > 0$ and case (i) or (ii) ensues. Also

$$`b^2 - 4ac` \;=\; 4[(Z - \hat{\alpha})^2\hat{\beta}^2 - (\hat{\beta}^2 - \hat{\sigma}^2 t^2/S_{xx})[(Z - \hat{\alpha})^2 - \hat{\sigma}^2 t^2(1 + 1/n)]$$

$$= \quad 4[(Z - \hat{\alpha})^2 \hat{\sigma}^2 t^2 / S_{xx} + (\hat{\beta}^2 - \hat{\sigma}^2 t^2 / S_{xx}) \hat{\sigma}^2 t^2 (1 + 1/n)],$$

and eqn (2.15) also makes '$b^2 > 4ac$', so that case (i) obtains.

This procedure for setting confidence sets is actually the generalized likelihood ratio method, $W(\hat{\xi}) - W(\xi) \le c$ where $W(\xi)$ is the log-likelihood maximized over $(\alpha, \beta, \sigma^2)$ for fixed ξ (see the next section for further details).

Approximate formulae may be obtained from the interval specified in case (i). Noting that $S_{xx} = \sum (x_i - \bar{x})^2 \stackrel{\text{def}}{=} n\sigma_x^2$ increases as n with probability 1 provided the x_i form a random sample from some distribution with finite variance, expanding formula (2.14) and retaining terms up to order $(1/n)$ the interval becomes:

$$[(Z - \hat{\alpha})/\hat{\beta}][1 + (\hat{\sigma}^2 t^2)/(\hat{\beta}^2 S_{xx})]$$
$$\pm (\hat{\sigma} t / \hat{\beta})[1 + 1/(2n) + \{(Z - \hat{\alpha})^2 + (\hat{\sigma}^2 t^2)\}/(2\hat{\beta}^2 S_{xx})].$$

If terms of order $(1/n)$ are ignored then the interval simply becomes

$$(Z - \hat{\alpha})/\hat{\beta} \pm \hat{\sigma} t / \hat{\beta}.$$

Further small sample and asymptotic results for the distribution of the two estimators, $\hat{\xi}$, $\check{\xi}$ are given in Section 2.6. An alternative asymptotic formula, arising from letting $\sigma \to 0$, could be derived and might be more accurate in small samples in precise calibrations. This form of *small sigma* asymptotics is used in Chapter 5. In some applications it is possible to take m genuinely independent replications of the future response at the particular unknown ξ. The next section presents the profile likelihood for controlled calibration allowing for this development. It leads to a simple estimator and confidence intervals based on the predictive distribution of $\bar{Z}$, the sample mean of the future Z, with a pooling of error variance from regression and future Z. No such simple generalization of the random calibration case is available, although relatively straightforward generalizations are possible, as discussed at the end of Section 2.5.

2.4   Profile likelihood with replications

In this section the profile likelihood is developed for the simple controlled calibration model with m replicated future Z at a single fixed unknown ξ. The profile log-likelihood for these $(n+m)$ observations is the log-likelihood maximized over the complete set of four parameters $(\alpha, \beta, \sigma^2, \xi)$, deducted from the log-likelihood maximized over the 'nuisance' parameters $(\alpha, \beta, \sigma^2)$ for fixed ξ, when these maximizing values will be functions of ξ. This may be plotted as a function of ξ and forms the basis of likelihood ratio tests and confidence intervals for ξ. Modifications of the profile likelihood would allow you to take more account of the uncertainty of the plugged-in maximum likelihood estimators (see Section 2.8).

We assume normality with models (2.1), (2.2), (2.6). For notational convenience let x and y of eqn (2.1) be centred, *post hoc* in the case of y, so that

$$\sum_1^n x_i = 0 \; ; \; \sum_1^n y_i = 0. \tag{2.16}$$

The centring of x allows a convenient reparametrization of eqns (2.1) and (2.2) to

$$Y_i = \mu + \beta x_i + \epsilon_i,$$

where $\mu = \alpha + \beta \bar{x}$, the mean of the Y-values with the old x-scale, and least-squares estimate $\hat{\mu} = 0$. The observation Z in eqn (2.6) may be scaled conformably, and ξ measured on the same scale as the centred x. The general case of m replicates in prediction gives

$$Z_l = \mu + \beta \xi + \epsilon_l, \quad l = 1, \dots, m \tag{2.17}$$

where ϵ_l are independent normal with zero mean and variance σ^2.

Firstly the maximized log-likelihood is evaluated when ξ is taken to be known. In this case the $(n+m)$ bivariate observations of models (2.2) and (2.17) may be combined as

$$Y_0 = 1\mu_0 + x_0\beta + \epsilon_0, \tag{2.18}$$

where $Y_0 = (Y_1, \dots, Y_n, Z_1, \dots, Z_m)$; $x_0 = (x_1 - a, \dots, x_n - a, \xi - a, \dots, \xi - a)$ with $a = m\xi/(n+m)$ and hence $\mu_0 = \mu + m\beta\xi/(n+m)$. Here a was chosen so that x_0 has mean zero, implementing a further centring, this time just for x. Both μ_0 and β are free to vary over the real line and are unknown. The log-likelihood of the $(n+m)$ independent observations is

$$-[(n+m)/2]log(2\pi\sigma_0^2) - \sum_1^{n+m} (y_{0i} - \mu_0 - x_{0i}\beta)^2/(2\sigma_0^2)$$

and is maximized at

$$\hat{\mu}_0 = \bar{y}_0, \quad \hat{\beta}_0 = \sum x_{0i} y_{0i} / \sum x_{0i}^2,$$

$$\hat{\sigma}_0^2 = \sum (y_{0i} - \hat{\mu}_0 - \hat{\beta}_0 x_{0i})^2/(n+m)$$

with maximized value

$$- [(n+m)/2]log(2\pi\hat{\sigma}_0^2) - (n+m)/2. \tag{2.19}$$

This is a function of ξ through $\hat{\sigma}_0^2$. From the orthogonality of 1 and x_0,

$$(n+m)\hat{\sigma}_0^2 = \sum(y_{0i} - \bar{y}_0)^2 - \sum(y_{0i} - \bar{y}_0)x_{0i}/\sum x_{0i}^2. \qquad (2.20)$$

It may be seen that

$$\sum(y_{0i} - \bar{y}_0)^2 = \sum_1^n y_i^2 + \sum_1^m z_i^2 - m^2\bar{z}^2/(n+m) \qquad (2.21)$$

$$\sum(y_{0i} - \bar{y}_0)x_{0i} = \sum_1^n y_i x_i + nm\xi\bar{z}/(n+m) \qquad (2.22)$$

$$\sum x_{0i}^2 = \sum_1^n x_i^2 + nm\xi^2/(n+m). \qquad (2.23)$$

The second term on the right-hand side of eqn (2.20) may be simplified; using eqns (2.22) and (2.23) it is equal to:

$$- \left[\sum x_i y_i + nm\bar{z}\xi/(n+m)\right]^2 / \left[\sum x_i^2 + nm\xi^2/(n+m)\right]$$
$$= -(\hat{\beta} + nm\bar{z}\xi g)^2/[gr(\xi)],$$

where $g = 1/\sum x_i^2$ and $r(\xi) = 1 + nmg\xi^2/(n+m)$. With further algebra,

$$= (\hat{\beta} + nm\bar{z}\xi g)^2[nm\xi^2/\{(n+m)r(\xi)\} - 1/g]$$
$$= (\hat{\beta}\xi - \bar{z} + \bar{z}r(\xi))^2 nm/\{(n+m)r(\xi)\} - [\hat{\beta} + nm\bar{z}\xi g/(n+m)]^2/g$$
$$= [(\hat{\beta}\xi - \bar{z})^2 + 2(\hat{\beta}\xi - \bar{z})\bar{z}r(\xi) + \bar{z}^2 r^2(\xi)]nm/[(n+m)r(\xi)]$$
$$\quad -[\hat{\beta}^2/g + 2\hat{\beta}nm\bar{z}\xi/(n+m) + n^2 m^2 \bar{z}^2 \xi^2 g/(n+m)^2]$$
$$= (\hat{\beta}\xi - \bar{z})^2 nm/[(n+m)r(\xi)] - \hat{\beta}^2/g - \bar{z}^2 nm/(n+m).$$

Substituting this back into eqn (2.20) and using eqn (2.21) gives

$$(n+m)\hat{\sigma}_0^2 = s_+^2 + (\hat{\beta}\xi - \bar{z})^2 nm/[(n+m)r(\xi)],$$

where s_+^2 is an error sum of squares from the regression of Y on x pooled with $\sum(z_i - \bar{z})^2$ on $n - 2 + (m - 1)$ degrees of freedom.

The log-likelihood is maximized when this is minimized, that is, at

$$\hat{\xi} = \bar{z}/\hat{\beta}, \qquad (2.24)$$

or, if Z is measured on an uncentred Y-scale, $\bar{z}$ is replaced by $\bar{z} - \bar{y}$ in the above formula. After exponentiation the profile likelihood is

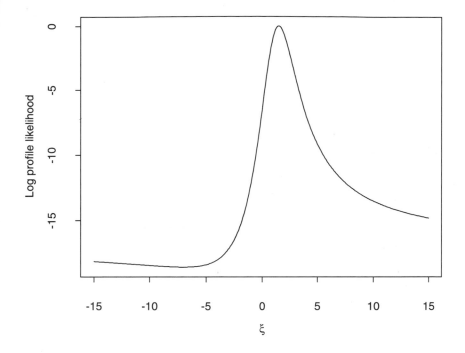

Fig. 2.8. Log-profile likelihood against ξ

$$\{c_{m,n}^2(\xi)/[c_{m,n}^2(\xi) + (\hat{\beta}\xi - \bar{z})^2/s_+^2]\}^{(m+n)/2} \qquad (2.25)$$

where

$$c_{m,n}^2(\xi) = (1/m + 1/n + \xi^2 g),$$

an extended notation from $c^2(x)$ of eqn (2.7). The profile likelihood is a monotone function of

$$(\bar{Z} - \hat{\beta}\xi)/[\hat{\sigma}(1/m + 1/n + \xi^2 g)^{1/2}],$$

with $(n + m - 3)\hat{\sigma}^2 = s_+^2$, and has a Student t-distribution on $n + m - 3$ degrees of freedom. This proves the earlier assertion that the profile likelihood is a function of the predictive pivotal Student statistic given by formula (2.8).

The profile likelihood given by (2.25) has the form shown (logged) in Fig. 2.8 (with $m = 1, n = 20, g = 0.1, \hat{\beta} = 2, z = 3, s_+^2 = 10$). It has two turning points, a maximum with the same sign as $\bar{z}$ at $\hat{\xi} = z/\hat{\beta} = 3/2$, and a minimum with the opposite sign at $-(1/m + 1/n)\hat{\beta}/(gz) = -7$. As $|\xi| \to \infty$ it tends to the positive

$$[1 + \hat{\beta}^2/(gs_+^2)]^{-(n+m)/2} = e^{-16.9}$$

from below at $-\infty$ and from above at $+\infty$. This failure to vanish at infinity is a reflection of a weak inferential position, the situation being changed only when either m or n tend to infinity. When $n \to \infty$, μ, β and σ^2 become known from the training data and the profile likelihood for ξ is as though it were normal with mean $\bar{z}/\hat{\beta}$ and variance $\hat{\sigma}^2/(m\hat{\beta}^2)$.

The above profile likelihood establishes the maximum likelihood estimator for controlled calibration and supports the method of setting confidence limits in Section 2.3. It provides a platform for taking account of uncertainty in the plugged-in maximum likelihood estimates of the nuisance parameters through a modified profile-likelihood. It will be seen to be closely comparable to an integrated likelihood or Bayesian posterior distribution for ξ, corresponding to vague prior knowledge.

2.5 Profile likelihood with some random X

Suppose (x_i, Y_i) for $i = 1, \ldots, n$ form regression data with controlled or designed x-values. However, separately there is a random sample of x-values, $X_1^*, \ldots, X_t^*$, distributed as normal with unknown mean and variance with ξ also being independently generated by this same mechanism. The profile likelihood for ξ solely from this marginal set of t values of x is easily shown by similar algebra to the above to be proportional to

$$\{1 + 1/t + (\xi - \bar{x}^*)^2 g^*\}^{-(t+1)/2} = \{c_{1,t}^2(\xi - \bar{x}^*)\}^{-(t+1)/2} \qquad (2.26)$$

where

$$g^* = \left\{ \sum_1^t (x_j^* - \bar{x}^*)^2 \right\}^{-1}.$$

Now eqn (2.26) multiplied by eqn (2.25),

$$c_{1,t}^2(\xi - \bar{x}^*)^{-(1+t)/2}\{1 + (\hat{\beta}\xi - \bar{z})^2/[s_+^2 c_{m,n}^2(\xi)]\}^{-(m+n)/2},$$

is proportional to the overall profile likelihood for ξ when the data consist of Y_i observed at fixed x_i, $i = 1, \ldots, n$, random X_j^*, $j = 1, \ldots, t$ and there are m replicates of Z at ξ. More interestingly, if $X_1^*, \ldots, X_t^*$ is in fact a subset of the x-values at which Y-values have been observed, this same product is still the overall profile likelihood. In the case of a bivariate normal distribution the parameters of the marginal distribution can be treated quite separately from those of the conditional distribution (a general property known as S-ancillarity, see Appendix B.2). If now $t = n$, so that in fact the situation is that of random or natural calibration w.l.o.g. taking $\sum x_i = 0$ the product gives the profile likelihood proportional to

$$c_{1,n}^2(\xi)^{-(n+1)/2}\{1 + (\hat{\beta}\xi - \bar{z})^2/(s_+^2 c_{m,n}^2(\xi))\}^{-(m+n)/2} \qquad (2.27)$$

which when $m = 1$ simplifies to

$$\{1 + 1/n + g\xi^2 + (\bar{z} - \hat{\beta}\xi)^2/s^2\}^{-(n+1)/2}.$$

On completing the square in ξ this may be reorganized as proportional to

$$\{gz^2/[s^2(g + \hat{\beta}^2/s^2) + (g + \hat{\beta}^2/s^2)(\xi - \check{\xi})^2]\}^{-(n+1)/2}$$

that is, it corresponds to the regression of X on y and has the Student predictive form of Section 2.2. Thus transition from the classical estimator for controlled calibration can be made in steps to the regression of X on Y, the inverse estimator, by injecting some random X distributed as ξ.

The approach above allowed the common unknown mean of the random X and ξ to be possibly different from that of the average of the fixed x-values. It offers an insight into the nature of the extra information that allows you to move from classical inference based on the regression of Y on x to the inverse estimator based on the regression of X on y. A parallel insight is given by the Bayesian analysis of Hoadley (1970), which will be considered in the extension to multivariate regression and calibration in Chapter 5. The maximum of the profile likelihood eqn (2.27) is the sensible, if complicated, generalization of the inverse estimator when replicates are present. In the notation of the regression of X on y as in Section 2.3, the simple estimator

$$\hat{\mu}_* + \hat{\beta}_* \bar{Z} \qquad (2.28)$$

is far from sensible. The regression slope coefficient of X on $\bar{z}$ is $[1 + \sigma^2/(m\beta^2\tau^2)]^{-1}$, which is a function of m, with τ^2 the variance of X and β, σ^2 parameters of the conditional distribution of Y given $X = x$ (see Appendix A.1). This naive estimator has a bias that persists even as the number of replicates goes to infinity.

The next section deals with the accuracy of the calibration estimators of ξ.

2.6 Estimation accuracy

During the 1960s a debate raged about the relative merits of $\hat{\xi}$ and $\check{\xi}$ given by eqns (2.11) and (2.10), respectively, as estimators of ξ. The debate largely focused on situations where the x_i are fixed in the training data. However, as pointed out by Krutckoff (1967), the mean of $\hat{\xi}$ does not exist and it has infinite mean-squared error. The inverse estimator $\check{\xi}$ is biased but has finite mean and variance. The estimator (2.11) is the ratio of two independent normal random variables, $Z/\sum a_i Y_i$, $a_i = (x_i - \bar{x})/S_{xx}$, with numerator expectation $\beta\xi$ and denominator expectation β (which would be Cauchy if their expectations were both zero). The mean and variance of the asymptotic distributions of $\hat{\xi}$ and $\check{\xi}$ are

$$E(\hat{\xi}) = \xi, \quad \mathrm{Var}(\hat{\xi}) = \sigma^2/\beta^2; \tag{2.29}$$

$$E(\check{\xi}) = \rho^2\xi + (1 - \rho^2)\bar{x}, \quad \mathrm{Var}(\check{\xi}) = \rho^4\sigma^2/\beta^2, \tag{2.30}$$

where

$$\rho^2 = \beta^2\sigma_x^2/(\sigma^2 + \beta^2\sigma_x^2),\ 0 \le \rho^2 \le 1.$$

With x centred, eqn (2.30) gives

$$E(\check{\xi}) = \rho^2\xi.$$

Shukla (1972) extends these formulae to an accuracy of order $(1/n)$ and Oman (1985) gives a series formula for the exact mean-squared error (and implicitly the mean and variance) of the inverse estimator $\check{\xi}$. Shukla (1972) also considers replications at ξ. Using estimator (2.24) reduces the variance in eqn (2.29) by a factor m. He also gives formulae for the mean and variance for the naive inverse estimator (2.28) whose bias persists even as m goes to infinity.

The mean-squared error is the sum of the variance and squared bias. It is hence easy to compare the mean-squared error of $\hat{\xi}$ and $\check{\xi}$ from the asymptotic formulae (2.29) and (2.30). The variance of $\check{\xi}$ is ρ^4 times that of $\hat{\xi}$ and $\rho^4 \ll 1$ if the 'signal-to-noise' ratio $\beta^2\sigma_x^2/\sigma^2$ is small. On the other hand $\check{\xi}$ is biased whereas $\hat{\xi}$ is not, asymptotically. The mean of $\check{\xi}$ is a weighted average of ξ (with weight ρ^2) and $\bar{x}$ (with weight $(1 - \rho^2)$). Its bias is small unless ξ is far from $\bar{x}$, the mean of the x-values in the training data. Broadly then, $\check{\xi}$ has smaller mean-squared error than $\hat{\xi}$ when ξ is close to $\bar{x}$, substantially so if the signal-to-noise ratio is small, otherwise the classical estimator $\hat{\xi}$ is preferable. This conclusion is supported by the results of the previous section where random X-values distributed like ξ when introduced move the maximum likelihood estimator from $\hat{\xi}$ towards $\check{\xi}$. In fact from the asymptotic formulae (2.29) and (2.30), in centred x-units,

$$\mathrm{MSE}(\hat{\xi}) - \mathrm{MSE}(\check{\xi}) = \sigma^2/\beta^2 - \rho^4\sigma^2/\beta^2 - (1 - \rho^2)^2\xi^2$$

and $\check{\xi}$ is better than $\hat{\xi}$ if and only if

$$\begin{aligned}
\xi^2 \ &< \ \{\sigma^2/\beta^2\}\{(1 + \rho^2)/(1 - \rho^2)\} \\
&= \ \{\sigma^2/\beta^2\}\{1 + 2\rho^2/(1 - \rho^2)\} \\
&= \ \sigma^2/\beta^2 + 2\sigma_x^2
\end{aligned}$$

$$= \quad \sigma_x^2(1/\rho^2 + 1). \tag{2.31}$$

In controlled regression the notion of regressing fixed values of x on Y is somewhat unsatisfactory from a sampling theory viewpoint. You might try to design the experiment so that in future the ξ would be well within the range of x-values, a sensible policy, but can you be sure of your success? Leaving that aside, it could be construed as a device to generate a better estimator without paying due care to wider inferential issues. It accommodates replications of future Z at a fixed ξ less readily. It utilizes information which might be argued to be only rightly available for Bayesian inference, and indeed the approach receives satisfactory backing through Theorem 5.2 and following discussion in Chapter 5. The non-existence of the expectation of $\hat{\xi}$ and its infinite variance and mean-squared error could be regarded as a proper reflection of the difficulties inherent in estimation from a sampling viewpoint. They may not be as serious as may at first be suspected. A Cauchy random variable has a well-defined median, and is unimodal with a central interval of 95 per cent probability content of finite length. However, its mean does not exist and it has infinite variance. In a sense, the infinite variance and mean-squared error of $\hat{\xi}$ are due to an unbounded quadratic loss function.

2.7 Generalized and weighted least-squares

A non-constant error variance may be suggested from *a priori* considerations, a graph of Y versus x, or residuals versus fitted values from ordinary least-squares. There may be suitable transformations of Y and possibly x which will give a simple mean relationship and a constant-variance additive error. If this does not seem possible then greater efficiency of estimation is obtained by a *weighted* least-squares analysis. The presence of correlation in the errors leads to a *generalized* least-squares analysis.

In general suppose that the variance of the n-vector of errors is given by

$$\text{Var}(\epsilon) = \sigma^2 W, \tag{2.32}$$

where W is an $n \times n$ known positive definite square symmetric matrix. Here $W = I_n$ reduces to the standard uncorrelated, constant variance case. If the form of W is ignored and one proceeds as if W were the identity matrix then least-squares estimates remain unbiased but are not generally fully efficient. That is, their variance matrix is greater than the generalized least-squares procedure to be described. In vector form the simple linear regression model is

$$Y = \mathcal{X}\theta + \epsilon, \tag{2.33}$$

with error variance matrix given by eqn (2.32), and $\theta = (\mu, \beta)$. Since W is known and symmetric we can always find a 'square-root' matrix U such that $U^T U = W$. Two possible candidates are the triangular Cholesky square-root or the symmetric square-root PDP^T based on the orthogonal diagonalization $P^T W P = D^2$ (see Appendix D, eqn (D.2)), where D^2 is the diagonal matrix of eigenvalues of W and P is an orthogonal matrix with eigenvectors as columns. The model eqn (2.33) may now be transformed to standard second-order assumptions. Since W is non-singular, $U^{-1} = PD^{-1}P^T$ exists, and multiplying across by this the model becomes

$$Y^* = \mathcal{X}^* \theta + \epsilon^* \tag{2.34}$$

with $Y^* = U^{-1}Y$, $\mathcal{X}^* = U^{-1}\mathcal{X}$, $\epsilon^* = U^{-1}\epsilon$, and $\mathrm{Var}(\epsilon^*) = \sigma^2 U^{-1} W U^{-T} = \sigma^2 I_n$. Thus standard least-squares formulae may be applied to model (2.34) producing fully efficient estimators

$$\hat{\theta} = (\mathcal{X}^{*T}\mathcal{X}^*)^{-1}\mathcal{X}^{*T}Y^* = (\mathcal{X}^T W^{-1}\mathcal{X})^{-1}\mathcal{X}^T W^{-1}Y \tag{2.35}$$

with covariance matrix

$$\sigma^2(\mathcal{X}^{*T}\mathcal{X}^*)^{-1} = \sigma^2(\mathcal{X}^T W^{-1}\mathcal{X})^{-1}.$$

The estimated intercept and slope parameters will be uncorrelated if this matrix is diagonal, that is, if the two columns of $\mathcal{X}$ are orthogonal with respect to the inner product $(u, v)_{W^{-1}} = u^T W^{-1} v$. The estimated error space is also orthogonal to the estimation space in the same sense and hence $(1, e)_{W^{-1}} = \sum \sum w^{ij} e_j = 0$ where $W^{-1} = (w^{ij})$ and w^{ij} is the ijth element of the inverse of W. In fact the estimates minimize the weighted sum of squares of errors $(\epsilon, \epsilon)_{W^{-1}}$.

Often the error covariance matrix factor W is not completely known. It may be that plots suggest a parametric specification with just a few unknown parameters. Gauss–Markov optimality of eqn (2.35) disappears when W is replaced by some estimate $\hat{W}$, except in large samples provided $\hat{W}$ is a consistent estimator of W. One standard technique is to iteratively reweight: by taking a trial $\hat{W}$ and estimating θ, then re-estimating W in the light of the new estimate of θ and cycling through until convergence obtains. See, for example, Green (1984) for a more extensive discussion.

Sometimes W is a function of x. Dependence of the error covariance matrix W on x adds some perhaps minor complexity to both prediction of a future response Z or inference about an unknown explanatory variable ξ when the response is observed. This marginally reinforces the transformation strategy that if at all possible dependence on x should be concentrated in the mean and not the variance. This strategy and its complement are applied to the bladder Example 1.4 at the end of this section. In passing note that if a random variable Y has mean μ and variance $g(\mu)$ then

$$h(y) = \int dy / \sqrt{g(y)}$$

approximately transforms to constant variance. For example, if $\mathrm{Var}(Y) = \sigma^2\mu$ then $\sqrt{Y}$ has a variance not depending on μ, at least approximately. Similarly if the standard deviation of Y is proportional to μ then the logarithm transformation is suggested (see Tukey (1957) for a taxonomy of strength of transformations). In the regression context it is harder to see when you should transform both Y and x, although prior knowledge of the structure of the problem may give hints. An entirely positive continuous random variable provides a hard-to-resist temptation to logarithmically transform. Generalized linear models (see for example McCullagh and Nelder (1988)) also provide an attractive alternative to transformation. The potential disadvantage of generalized linear models is that they are conceived largely on prior grounds whereas, in contrast, the potential weakness of transformations is that they are based predominantly on empirical observation.

One may be interested in prediction of a future m-vector Z under the model

$$Z = \mathcal{X}_1\theta + \epsilon_1 \tag{2.36}$$

where $\mathcal{X}_1$ is $m \times 2$ and ϵ_1 has covariance matrix $\sigma^2 W_1$, with W_1 known. A multivariate Student predictive distribution can be derived for Z. As a first step in its derivation note that for σ^2 known, the predictive distribution of $Z - \mathcal{X}_1\hat{\theta}$ is m-variate normal with zero vector mean and covariance matrix

$$\{W_1 + \mathcal{X}_1(\mathcal{X}^T W^{-1}\mathcal{X})^{-1}\mathcal{X}_1^T\}\sigma^2.$$

The weighted least-squares fitted line in the calibration experiment provides a standard unbiased estimator of σ^2, from the residual sum of squares divided by degrees of freedom,

$$Y^{*T}(I - H^*)Y^* / (n - 2), \tag{2.37}$$

with the 'hat' matrix

$$H^* = \mathcal{X}^*(\mathcal{X}^{*T}\mathcal{X}^{*T})^{-1}\mathcal{X}^* = W^{-1/2}\mathcal{X}(\mathcal{X}^T W^{-1}\mathcal{X})\mathcal{X}^T W^{-1/2}.$$

Suppose now that the m values of Z in model (2.36) are observed at a single unknown value of the explanatory variable, ξ, and that you are interested in estimating this unknown. If the mean in model (2.36) is constant then an appropriately weighted mean of the m values of Z is sufficient for ξ, and the $m - 1$ contrasts serve to estimate σ^2. In short the inference for ξ as in Section 2.3 follows based on the pivotal Student t-statistic on $n + m - 3$ degrees of freedom,

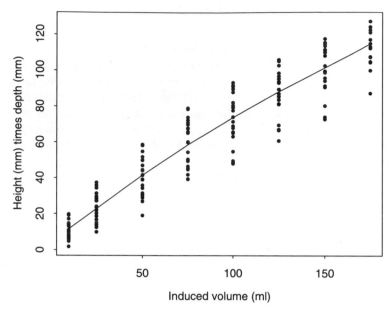

Fig. 2.9. Bladder data with locally weighted regression fit

$$(\bar{Z}^* - \hat{\mu} - \hat{\beta}\xi)/[\hat{\sigma}\sqrt{\{a_{W_1} + (1,\xi)(\mathcal{X}^T W^{-1} \mathcal{X})^{-1}(1,\xi)^T\}}],$$

with $\bar{Z}^* = 1^T W_1^{-1/2} Z / 1^T W_1^{-1/2} 1$, $a_{W_1} = m/(1^T W_1^{-1/2} 1)^2$ and $\hat{\sigma}^2$ is pooled from calibration and prediction.

When the covariance matrices are functions of x or in the case of W_1, ξ, then a_{W_1} is a function of ξ further modifying the pivotal t-statistic as a function of ξ. One relatively simple illustrative case is if the errors $\{\epsilon_i\}$ may be taken to be uncorrelated but they have variance $\sigma^2 w(x_i)$, with $w(x_i)$ a known function of x_i. Here the matrix W is diagonal with ith diagonal element $w(x_i)$. The weighted least-squares estimation proceeds as above and since each of the m-replicates of Z now have the same variance $\sigma^2 w(\xi)$,

$$a_{W_1} = w(\xi)/m.$$

The profile likelihood is a function of the pivotal t-statistic as in the unweighted case. The profile likelihood calculation involves further removal of nuisance parameters if W and W_1 are not completely known but parametrically specified. Alternatively, consistent estimators of these nuisance parameters in the covariance function allow you to proceed as above but now with an asymptotically valid analysis.

At this point it is appropriate to say a little more about the bladder data of Example 1.4. The height by depth measurements are for a series

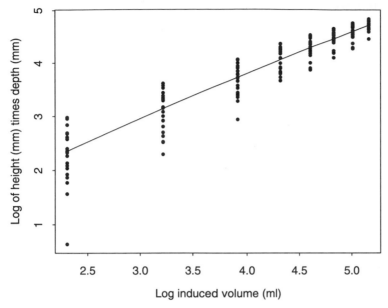

Fig. 2.10.   Bladder data, logged, and locally weighted regression fit

of fluid volumes for each of 23 women. Thus the data have a longitudinal or repeated measurement aspect. The plot of *hd* against volume for
a randomly chosen woman, say the first, is of a systematic form, similar
to that of the average over recorded data of the 23 women in Fig. 1.4. A
locally weighted least-squares analysis, Fig. 2.9, using the method of Cleveland (1979) and programmed as the *lowess* function in S (Becker *et al.*,
1988) gives much the same picture. The curves do vary across the women,
and the average curve is somewhat curved. Any analysis which has preten-
sions to extracting the maximum information from the data might envisage
correlated errors within a subject. The *hd* measurement has dimension of
(length)^2 and volume has dimension (length)^3. One would hardly imagine
a linear relationship. Simplistically the *hd* measurements raised to the 3/2
power might be proportional to volume, or logged *hd* might be linear in log
volume. Indeed this seems from Fig. 2.10 to be so. However, the variance
is now far from constant, changing with log volume. There is the choice of
a linear but weighted analysis on this logged scale or non-linear mean relationship (perhaps quadratic) but with constant variance (approximately)
for the unlogged variables. In the former case, at least for calibration, the
variance needs to be modelled as a function of logged volume. The standard deviation of logged *hd* is about linear in log volume over the range
of data. Both methods should lead to similar predictive intervals, at least
within the range of the data.

2.8   Bibliography

The controversy over which estimator to use for prediction of the explanatory variable in simple regression was fuelled by Krutchkoff (1967) and answered by Berkson (1969) and Williams (1969). Bayesian resolutions of the controversy were offered by Hoadley (1970), Aitchison and Dunsmore (1975) and Hunter and Lamboy (1981). Lwin and Maritz (1980) combine the conditional distribution of Y for given $X = x$ with an empirical estimator of the marginal distribution of random X.

Minder and Whitney (1974) give a marginal likelihood approach, and the profile likelihood inference presented here is derived from Brown and Sundberg (1987, 1989); see also Harding (1986) and Aitkin *et al.* (1989). For a discussion of modifications to the profile likelihood idea see Cox (1988). Locally weighted regression points to one method of approaching non-linearity; a topic of Chapter 7.

3
Multiple regression and calibration

3.1 Least-squares

The previous chapter looked at the relationship between two variables: the response Y and the explanatory variable x, and with n observations developed model fitting by least-squares. With a response and explanatory variable pair, it emphasized both the prediction of the future Y, denoted Z, for given x; and estimation and confidence intervals for unknown x, denoted ξ, for given Z. In this chapter these ideas are extended to models where more than one variable is available to explain Y. This chapter concentrates on least-squares methods whereas the following chapter looks at less-standard *shrinkage* methods. These are capable of dealing sensibly with both non-singular and singular specifications of the linear model.

For the linear regression model a single response variable Y is such that for given p-vector $x = (x_1, \ldots, x_p)$, the conditional expected value of the response, $E(Y|x)$, is

$$E(Y|x) = \alpha + \beta^T x. \tag{3.1}$$

Here $x = (x_1, \ldots, x_p)^T$ may be observed realizations of p distinct random variables, p constructed values derived from $p' \leq p$ random variables, or they may be p prespecified (designed) constants, or even a mixture of these 'extreme' cases. Since the model does not always refer to random explanatory variables treated conditionally, explicit mention of the conditioning argument is often dropped and $E(Y|x)$ simply replaced by $E(Y)$. Model (3.1) is both linear in the carriers x and the parameters, but it is the linearity in the $p+1$ unknown parameters, α and β, which renders the regression *linear*. Since the p carriers may be non-linear functions of originating variables the relationship may be far from linear in these variables. For example the regression $E(Y|x) = \alpha + \beta_1 x + \beta_2 x^2$, quadratic in the single variable x, is nonetheless a multiple linear regression. This greatly extends the usefulness of the linear regression methodology, and for a variety of linearizable forms see, for example, Cox and Snell (1981, p. 29–33). However, the non-linearity in x in this quadratic example returns as model non-linearity in the estimation of a future x from a given future Y.

Superimposed on the conditional expectation of eqn (3.1) is an additive error, assumed to be uncorrelated with the x-variables and having variance

σ^2, giving the model

$$Y = E(Y|x) + \epsilon.$$

Typically to be able to fit such a relationship n uncorrelated observations $Y_1, \ldots, Y_n$ are observed at realizations or specifications $x_1, \ldots, x_n$ of the p carriers. The full rank case requires that the $n \times (p+1)$ model matrix $\mathcal{X}$ with ith row $(1, x_i^T), i = 1, \ldots, n$, is such that $n \geq (p+1)$ and is of full rank $(p+1)$. Relaxation of these assumptions, essentially concerned with the estimability of β, is left to the following chapter.

The n-observation linear regression model is thus

$$Y = \mathcal{X}\theta + \epsilon \tag{3.2}$$

with $\theta = (\alpha, \beta^T)^T$. Under second-order assumptions, the error has mean the zero vector, and covariance matrix $\sigma^2 I_n$. It is also assumed uncorrelated with the x-variables. Such an assumption is often not stated but is crucial in examining difficulties with least-squares estimation when the x-variables are only observed with superimposed errors, the case of errors in variables, touched on in Chapter 7. Occasionally it is desirable to fit a relationship without a constant, so that $\mathcal{X}$ is $n \times p$ and $\theta = \beta$. However, such occasions are rarer than might be imagined. Insistence that the line passes through the origin when the bulk of the data are far from the origin constitutes a rather strong assumption. Even though for theoretical reasons it may be anticipated to pass through the origin, the relationship may not be perfectly linear over the entire region despite being locally linear enough within the data. Extrapolation of the locally linear relation will then not pass through the origin.

A least-squares estimator minimizes the residual sum of squares

$$\sum_{i=1}^{n} (y_i - \alpha - \beta_1 x_{i1} - \cdots - \beta_p x_{ip})^2$$

and is a solution of the 'normal' equations,

$$\mathcal{X}^T \mathcal{X} \theta = \mathcal{X}^T y. \tag{3.3}$$

A linear combination of parameters, $c^T \theta$, is estimable if the constant $(p+1) \times 1$ vector c lies in the space spanned by rows of X. In the full rank case all $(p+1)$ parameters are uniquely estimable by

$$\hat{\theta} = (\mathcal{X}^T \mathcal{X})^{-1} \mathcal{X}^T y \tag{3.4}$$

and for any vector of constants c, $c^T \hat{\theta}$ is unbiased for $c^T \theta$ and has amongst unbiased estimators linear in y the minimum variance,

$$c^T(\mathcal{X}^T\mathcal{X})^{-1}c\sigma^2. \tag{3.5}$$

This constitutes a statement of the Gauss–Markov theorem.

The fitted residuals, $e = y - \mathcal{X}\hat{\theta}$, are orthogonal to the columns of $\mathcal{X}$. The residuals are linear in y, since

$$e = \{I_n - \mathcal{X}(\mathcal{X}^T\mathcal{X})^{-1}\mathcal{X}^T\}y = (I_n - H)y = Py$$

where H is the 'hat' projection matrix (see Appendix D), and P is a projection matrix to the $n-p-1$ dimensional space orthogonal to the $(p+1)$-space spanned by the columns of $\mathcal{X}$. Normality plus the second-order assumptions imply the residual sum of squares, $e^T e$, is distributed as a chi-squared random variable scaled by σ^2, and is independent of $\hat{\theta}$. Thus

$$\hat{\sigma}^2 = e^T e/(n-p-1)$$

is unbiased for σ^2. It is also a function of the sufficient statistics and may be calculated directly from

$$e^T e = y^T P^2 y = y^T P y = y^T y - \hat{\theta}^T \mathcal{X}^T \mathcal{X}\hat{\theta} = y^T y - \hat{\theta}^T \mathcal{X}^T y.$$

For numerical purposes, however, it is better not to 'square' the matrix $\mathcal{X}$ in the first place as in the normal equations (3.3). It is possible to do a so-called QR-decomposition of $\mathcal{X}$ where Q is orthogonal and R upper triangular, perhaps directly by a series of Householder elementary orthogonal transformations to reduce the matrix $\mathcal{X}$ and correspondingly transform and reduce y. The least-squares estimates are then directly obtained by successively solving the $p + 1$ triangular set of equations, $R\theta = Q^T y$ from the bottom for one unknown variable at a time (see, for example, Thisted (1988, p. 63–65)).

The plots of residuals, either with or without deletion, suggested in Section 2.1 are available for checking for outliers, departures from assumptions in the model, and possible model refinements. For a notable example of graphical detective work see Denby and Pregibon (1987).

3.1.1   ROLE OF ORTHOGONALITY

Suppose the model matrix $\mathcal{X}$ is partitioned into a set of $(p + 1 - r)$ and r variables, $\mathcal{X} = (X_1, X_2)$. Then the $n \times (p + 1 - r)$ matrix X_1 and $n \times r$ matrix X_2 are orthogonal if any column of X_1 is orthogonal to any column of X_2, that is, if $X_1^T X_2 = 0$, a $(p + 1 - r) \times r$ matrix of zeros. Thus the spaces spanned by column vectors from each matrix are orthogonal. Since the covariance matrix of $\hat{\theta}$ is

$$\sigma^2(\mathcal{X}^T\mathcal{X})^{-1}$$

in this case of orthogonality this becomes the product of σ^2 and

$$\begin{pmatrix} (X_1^T X_1)^{-1} & 0 \\ 0 & (X_2^T X_2)^{-1} \end{pmatrix}.$$

Partitioning the parameter conformably, $\theta = (\theta_1, \theta_2)$, then under normality estimators $\hat{\theta}_1, \hat{\theta}_2$ in the model (3.2) are independently distributed. Such independence can always be made between the intercept estimator $\hat{\alpha}$ and the slope coefficients $\hat{\beta}$ just by centring all the x-variables, that is,

$$\sum_i x_{ij} = 0, \ j = 1, \ldots, p. \tag{3.6}$$

This is also good numerical practice and will be assumed in the sequel. The model may then be presented in a slightly different parametrization,

$$Y = \mu 1_n + X\beta + \epsilon, \tag{3.7}$$

with $\mu = \alpha + \sum \beta_j \bar{x}_j$ now the mean of Y, in terms of the old means $\bar{x}_j$ of x_j, and $n \times p$ centred model matrix X. Orthogonality is also built into classical design of experiments.

3.2   Testing and variable selection

In the previous section the fitted residual sum of squares was shown to be

$$Y^T Y - \hat{\theta}^T \mathcal{X}^T Y = \sum (Y_i - \hat{\mu} - \hat{\beta}^T x_i)^2 = s^2.$$

By the remark on orthogonality of $\hat{\mu}$ and $\hat{\beta}$ after centring the x-variables, this may be simplified to

$$y^T y - \hat{\beta}^T X^T X \hat{\beta},$$

where y as well as X are centred.

The mean corrected total sum of squares $y^T y$ is reduced by the sum of squares explained by the the x-variables, $\hat{\beta}^T X^T X \hat{\beta}$. The explained sum of squares is distributed independently of the residual sum of squares. The analysis of variance table acts as an *aide memoire* to testing the hypothesis that $\beta = 0$.

Under normality and $\beta = 0$ the ratio of regression mean-square to residual mean-square has a central F-distribution on $(p, n - p - 1)$ degrees of freedom. Away from $\beta = 0$ the numerator sum of squares has a non-central chi-squared scaled by σ^2. A useful summary of the explanation of

Table 3.1. Anova table for testing $\beta = 0$

Source	SS	Df	MS	F
Due to β	$\hat{\beta}^T X^T X \hat{\beta} = b$	p	$b/p = d$	
residual	$y^T y - \hat{\beta}^T X^T X \hat{\beta} = e$	$\nu = n - p - 1$	$e/\nu = \hat{\sigma}^2$	$d/\hat{\sigma}^2$
Total	$y^T y$	$n - 1$		

variation is the multiple correlation, R, the sample correlation between y and $\hat{y}$. As in simple linear regression R^2 is the proportion of variation in y explained.

To test a null hypothesis $K\beta = 0$, where K is a specified known matrix of constants of rank k, then in essence two models are fitted; the above unrestricted model and a model restricted by $K\beta = 0$, with $r = p - k$ variables. Under normality the reduction in sum of squares in moving from the restricted to the full model has a chi-squared distribution on k degrees of freedom (scaled by σ^2) and is distributed independently of the residual sum of squares from the full model. An F-test may be constructed by taking the ratio of the mean-squares. If $k = 1$ and one wishes simply to test say $\beta_p = 0$, then the F-test on $(1, n - p - 1)$ degrees of freedom is identical to a Student two-sided t-test of the same hypothesis based on the standard error given in eqn (3.5) with $c = (0, \ldots, 0, 1)^T$. The Student test may be preferred as it gives the direction in addition to the size of departure from the null hypothesis. A third equivalent testing procedure is to regress through the origin the residuals from the regression of y on $(x_1, \ldots, x_{p-1})$, on the residuals from regressing x_p on the same regressors. The plot of the two sets of residuals, an added variable plot, is an informative diagnostic for the usefulness of the added variable, x_p. The standard t-test of zero slope of the line through the origin provides an exactly equivalent test to the F-test and t-test above.

Automated methods of model fitting which avoid all 2^p submodels are popular in computer packages. Most favoured are forward selection, backward selection, and stepwise selection. Each of these with their specific protocols for inclusion and/or deletion define a selection strategy. Except where the purpose is prediction of the response from explanatory variables generated as in the training data, and the physical interpretation of the relationship is unimportant, such methods should be used with caution. When the explanatory variables are highly correlated, different sets of x-variables can act as surrogates for one another and any single definitive relation is suspect. A variety of quite distinct models may fit the response reasonably well. An extensive automated sifting process should (1) admit that other models may be appropriate, (2) validate a chosen reduced model on further datasets since spurious models may be generated by the extent of the sifting process.

Perhaps the best of the above selection procedures is stepwise selection, which builds up a set of variables successively and allows earlier chosen variables to be discarded. Variables are introduced or discarded one at a time. Suppose the current subset of r variables, called set S_r, has a significant F-value at a chosen level. The following steps or variations on these are initiated.

1. Deletion of variables. Is the slope coefficient for any one of the variables in S_r not significantly different from zero? This can be seen by scanning the r Student ratios for the model with all r variables fitted. If more than one variable has an insignificant coefficient then the least significant variable is deleted. Further checking for insignificant coefficients is possible for the model with $(r-1)$ variables fitted.

2. Addition of a variable. Search all variables in the complement of S_r for one which increases the explained sum of squares the most. If this increase is statistically significant then include this variable in S_r.

3. Go to step 1 and repeat the cycle until further deletions or additions are not significant.

Kennedy and Bancroft (1971) suggest a 25 per cent point of the F-distribution for addition and a 10 per cent value for deletion. Such an algorithm looks through a rather restricted set of the 2^p possible sets. All subsets algorithms are computationally feasible for problems with relatively few variables, an increasingly rare situation these days. Branch and bound algorithms allow one to search much less than 2^p regressions, but are still computationally demanding for problems with many variables.

A guard against overfitting is provided by Mallows C_p. This ought here to be called C_r, but the terminology is well ingrained and for the following description p temporarily denotes the dimension of a submodel. C_p penalizes models with a large number of parameters. Mallows (1973) considered classes of estimators of β, linear in Y, but typically biased, and looked for unbiased estimators of their prediction mean-squared error. A rationale is that the mean-squared error is the sum of the variance and squared bias; the use of biased estimators reduces the former whilst increasing the latter. It is desirable to avoid a model so large that the bias takes over. Formally his criterion for choice of model is to minimize over p

$$C_p = RSS/\hat{\sigma}^2 - (n - 2p).$$

The first term decreases with larger models and is counter-balanced by the increasing second term. Here $\hat{\sigma}^2$ is the least-squares estimate from the largest model envisaged. More often C_p is used as a graphical tool, plotting C_p against p for different sized models. Perhaps the greatest value of the device, quoting Mallows (1973), 'is that it helps the statistician to examine some aspects of structure in her data and helps her to recognize ambiguities that confront her.' Mallows C_p generalizes in non-normal models to Akaike's information criterion (AIC) (see Akaike (1974)). A Bayesian variant (BIC) due to Schwarz (1978) penalizes twice the log-likelihood ratio by $p \log(n)$ rather than the $2p$ of Akaike. Utilizing Laplace's method, Kass and Vaidyanathan (1992) show that this can be interpreted as a first approximation to the Bayes factor for comparing two models.

The use of prediction as a tool for validation is explored in the next section.

3.3 Least-squares prediction

The Gauss–Markov theory of optimal least-squares estimation may be extended to least-squares prediction. Suppose you wish to make inference statements about a future response, denoted Z, at a prespecified value x of the p explanatory variables. If Z behaves like Y in the training data at x measured on the scale of the centred model of eqn (3.7), and has error with the same variance σ^2, then

$$Z = \mu + \beta^T x + \epsilon. \tag{3.8}$$

Suppose $d(x)$ $(n \times 1)$ is a vector function of x to be specified, such that $\hat{Z} = d(x)^T Y$, linear in $Y = (Y_1, \ldots, Y_n)^T$ for any x, is demanded to be prediction unbiased:

$$E(\hat{Z}) = E(Z), \tag{3.9}$$

and has minimum mean-squared error. Choosing $d(x)$ such that

$$E(Z - \hat{Z})^2$$

is minimized subject to constraint eqn (3.9) for all x, gives the best linear predictor,

$$\hat{Z} = \hat{\mu} + \hat{\beta}^T x$$

where not surprisingly $\hat{\mu}, \hat{\beta}$ are least-squares estimators. The minimized mean-squared error of prediction is

$$\sigma^2 \{ 1 + (1, x^T)(\mathcal{X}^T \mathcal{X})^{-1}(1, x^T)^T \} \overset{\text{def}}{=} \sigma^2 c^2(x). \tag{3.10}$$

This is easily generalizable to prediction at a set of m future Z, as in the weighted least-squares section of the previous chapter. With σ^2 estimated by $\hat{\sigma}^2$ then

$$(Z - \hat{\mu} - \hat{\beta}^T x)/\{\hat{\sigma} c(x)\}$$

has a Student t-distribution on $\nu = (n - p - 1)$ degrees of freedom. This sampling distribution is relevant for repetitions of $\{Z, Y_1, \ldots, Y_n\}$ for fixed

$\{x, x_1, \ldots, x_n\}$. The Student distribution does not, however, involve $\{x, x_1, \ldots, x_n\}$ and the distribution is also valid unconditionally. This distribution also corresponds to a Bayes predictive distribution for Z conditional on fixed $\{y_1, \ldots, y_n, x, x_1, \ldots, x_n\}$ for a particular 'vague' prior distribution (see Chapter 5). A $100(1-\gamma)$ per cent prediction interval for Z is then

$$\hat{\mu} + \hat{\beta}^T x - t_\nu^*(\gamma)\hat{\sigma}c(x) \leq Z \leq \hat{\mu} + \hat{\beta}^T x + t_\nu^*(\gamma)\hat{\sigma}c(x) \qquad (3.11)$$

where $t_\nu^*(\gamma)$ is the tabulated upper $100(\gamma/2)$ percentage point of the Student t-distribution on ν degrees of freedom. This predictive distribution provides the natural vehicle for predicting a future explanatory variable when that explanatory variable has the same random mechanism generating it in both calibration and prediction. On the other hand the next section describes estimation when such natural randomness cannot be assumed, as in controlled calibration.

3.4   Controlled calibration and polynomial regression

Suppose the relationship between the response and explanatory variables has been calibrated and a new response Z is observed corresponding to an unknown p-vector ξ of explanatory variables. The vector ξ is then indeterminately specified by Z. The p-components of ξ are free to vary in a $(p-1)$ dimensional subspace specified by $\hat{\mu} + \hat{\beta}^T\xi = Z$. This indeterminacy renders the case of little practical interest. The case comes to life, however, if the mean of Z is only a one-dimensional function of ξ. This is so, for example, in polynomial regression when say $(\xi, \xi^2, \ldots, \xi^p)$, ξ scalar, replaces the vector ξ. For this uncentred formulation, $h(\xi) = \hat{\alpha} + \sum \hat{\beta}_i \xi^i$, the Student pivot (2.8) could be used as a basis for confidence regions. A $100(1-\gamma)$ per cent confidence set for the scalar ξ is then those ξ such that

$$|Z - h(\xi)|/\{\hat{\sigma}c(\xi)\} \leq t_\nu^*(\gamma), \qquad (3.12)$$

with $c(\xi)$ from eqn (3.10). This set will be an interval typically if the values of ξ are restricted *a priori*. Whilst it is exact it should not be used uncritically. There may, for example, be no single value of ξ such that $Z = h(\xi)$. Moreover as the minimum over ξ of the left-hand side of eqn (3.12) approaches the right-hand side the region will shrink to a point. It is empty when this minimum is greater than the critical tabulated Student value. In practical terms this is acceptable only if it is regarded as a signal that something has gone wrong with the calibration. A much better solution is offered in the next section by a profile likelihood approach, although even here there is a need for a warning to the user that the Z is assumed generated from exactly the same mechanism as the calibrating Y. Further values of Z at the same ξ can at least verify the constancy of error variance.

3.5   Profile likelihood

We consider controlled calibration. To derive the profile likelihood for ξ in multiple linear regression, the approach for simple linear regression given in Section 2.4 may be used. Assume model (3.7) for the training data, and an m-replicates version of the prediction model (3.8),

$$Z_l = \mu + \beta^T \xi + \epsilon_l, \quad l = 1, \ldots m. \tag{3.13}$$

All errors are assumed independent normal with zero mean and variance σ^2. The profile likelihood of the ξ is

$$\{c_{m,n}^2(\xi)/[c_{m,n}^2(\xi) + (\hat{\beta}^T \xi - \bar{z})^2/s_+^2]\}^{(m+n)/2} \tag{3.14}$$

where
$$c_{m,n}^2(\xi) = (1/m + 1/n + \xi^T G \xi), \tag{3.15}$$

and $G = (X^T X)^{-1}$, $s_+^2 = s^2 + \sum(z_l - \bar{z})^2$. This is a vector ξ version of the profile likelihood and extends the scalar ξ version given by formula (2.25). Profile likelihood (3.14) is a monotone function of the predictive pivotal,

$$(\bar{Z} - \hat{\beta}^T \xi)/\{\hat{\sigma} c_{m,n}(\xi)\}, \tag{3.16}$$

with $(n + m - p - 2)\hat{\sigma}^2 = s_+^2$, which has a Student t-distribution on $\nu = (n + m - p - 2)$ degrees of freedom, given ξ. When ξ is completely free to vary there is a $(p - 1)$ dimensional hyperplane of possible maximum likelihood estimators, namely those $\hat{\xi}$ which minimize $(\hat{\beta}\xi - \bar{z})^2$. Here the calibration is too poorly specified to be of interest. However, formula (3.14) is proportional to the profile likelihood even when the mean of Z is a non-linear function of scalar ξ whilst linear in the regression parameters. For example, the case of polynomial regression may have $E(Z) = \mu + \beta_1 h_1(\xi) + \ldots, \beta_p h_p(\xi) = h(\xi)$, and this is constructed from orthogonal polynomials, allowing all the regression coefficients to be independently estimated. Figure 3.1 plots a quadratic $h(\xi)$ against ξ. Illustrated are two possible values of $\bar{z}$, one, $\bar{z} = 0.5$, cuts the curve in two values of ξ and the other, $\bar{z} = 2$, does not cut it at all. In the former case there are two values of ξ maximizing (3.14) (with $h(\xi)$ replacing $\beta^T \xi$), whereas in the latter case there is only one value. Thus care should be exercised to see that the estimators and likelihood ratio confidence intervals are within the range of x-values fitted. If one can assume that the relationship is monotone over the range of possible values of ξ, and that the error variance is small so that $\bar{Z}$ does not correspond to ξ outside this range, then there will be a unique maximum likelihood estimator. An approximate confidence interval

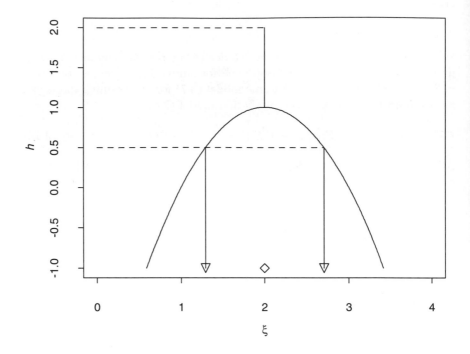

Fig. 3.1. Quadratic regression and possible estimators of ξ

from the the profile likelihood, which is a function of the pivotal predictive Student ratio (3.16), may be obtained from a Taylor series expansion about the solution $\hat{\xi}$ to $h(\xi) = \bar{z}$. To a first approximation the numerator of formula (3.16) is $|(\hat{\xi} - \xi)h'(\hat{\xi})|$ and the denominator may be just evaluated at $\hat{\xi}$. An approximate confidence interval for the scalar ξ in this polynomial case is

$$\hat{\xi} - t_\nu^*(\gamma/2)\hat{\sigma}c(\hat{\xi})/|h'(\hat{\xi})| \leq \xi \leq \hat{\xi} + t_\nu^*(\gamma/2)\hat{\sigma}c(\hat{\xi})/|h'(\hat{\xi})|$$

with

$$c^2(\hat{\xi}) = 1/m + 1/n + \sum_{j=1}^{p} h_j^2(\hat{\xi}) / \sum_{i=1}^{n} h_j^2(x_i),$$

where $h_j(.)$ is an orthogonal polynomial of degree j, $j = 1, \ldots, p$, so that $\sum_{i=1}^{n} h_j(x_i)h_{j'}(x_i) = 0, j \neq j'$, and orthogonal to the unit vector.

3.6   Bibliography

This chapter covers much that is standard in multiple regression and is covered in a variety of textbooks. Williams (1959) provides an early reference with an emphasis on calibration problems. For geometric insights into least-squares see Scheffé (1959). The profile likelihood approach to calibration comes from Brown and Sundberg (1987). Lundberg and de Maré (1980) give approximate calibration confidence intervals for polynomial regression. An alternative approach to calibration, both linear and non-linear, is provided by Scheffé (1973). He is concerned about the overall unconditional error rates in repeated use of a fitted calibration curve. It is arguable that such rates are not relevant to the specification of uncertainty for a specific prediction. In practice the repeated use intervals are very wide for typical global confidence levels.

Smoother alternatives to fitting a high-order polynomial are provided by regression splines (see for example de Boor (1978) for a definitive text). In this a knot sequence and the order k of the polynomial between each knot is specified, and the spline is fitted by least-squares. A spline of order k has continuous derivatives up to order $(k-1)$. Cubic splines are particularly popular. The method does require careful choice of the number and position of knots. The approach, akin to a computational version of a french curve, is unlikely to be underpinned by a substantive interpretation.

4
Regularized multiple regression

4.1 Introduction

The Gauss–Markov theorem lends considerable weight to the use of least-squares as a method of fitting a linear parametric model. It provides 'best' estimators. It does, however, have a very restricted notion of what is best. Firstly it confines attention to estimators linear in the response Y. Secondly it looks only at unbiased estimators. Thirdly it minimizes mean-squared error, that is, it is concerned with expected squared error or quadratic loss: in the scalar case it uses $E(\hat{\theta} - \theta)^2$ as a measure of how good $\hat{\theta}$ is as an estimator of θ. Even retaining quadratic loss, there may be estimators which are either biased or non-linear in Y or both, which have an average performance far better than least-squares over much of the relevant parameter space. Such estimators may be introduced in a variety of ways. They are often thought to be particularly apt when the model matrix, X, becomes singular or rather near to singular. They are sometimes called regularized estimators and at other times shrinkage estimators, due to the reduced length of the estimated regression parameter vector compared to that of least-squares. The term regularized emanates from the method of regularization in approximation theory literature. Introducing a Lagrange multiplier, λ, this amounts to minimization of

$$||Y - \mu - X\beta||^2 + \lambda||\beta||^2, \tag{4.1}$$

or its Hilbert space generalization. Their use as estimators or predictors comprises a large literature; a much smaller literature addresses wider inferential questions. In the model (3.7) assumed here, where columns of X have mean zero, generally $\hat{\mu} = \bar{y}$, as in standard least-squares estimation.

In this chapter a few regularization methods commonly used are described in some detail: ridge regression (RR), principal components regression (PCR), and partial least-squares regression (PLSR). Many of these regularized estimators may be put into the form

$$\hat{\beta}_R = \mathcal{G}X^TY \tag{4.2}$$

or

$$\hat{Y}_R = \hat{\mu} + HY \qquad (4.3)$$

where $H = XGX^T$. Here G approximates the 'inverse' of X^TX in various ways. This approximation may involve the vector Y when $\hat{\beta}_R$ becomes non-linear in Y. Ridge regression takes $G = (X^TX + \hat{k}I_p)^{-1}$, where $\hat{k}$ is the ridge constant estimated in some way from the data, often involving Y. Principal components regression starts from a spectral decomposition (see Appendix D),

$$X^TX = \sum_{j=1}^{p} \lambda_j v_j v_j^T$$

and takes

$$G = \sum_{j \in S_\omega} (1/\lambda_j) v_j v_j^T, \qquad (4.4)$$

where S_ω is a subset of $\omega \le \min(p, n-1)$ indices of the p variables chosen by one of two methods:

(a) explanation of variance, that is, based on eigenvalues of X^TX;

(b) based on those eigenvectors for which the correlation with Y is high.

Case (a) constitutes traditional PCR, whereas case (b) is a hybrid version that involves Y in the specification of G. The third regularization method, partial least-squares regression, inverts X^TX by what amounts to a conjugate gradient algorithm for matrix inversion, where the number of conjugate directions is specified by the data.

  If the number of observations is greater than the number of variables and X is of full rank p then prediction of Y from ordinary least-squares is not changed by rescaling and linearly transforming the p carrier variables. This is not true of RR, PCR, or PLS except in the special cases where they give ordinary least-squares or where only orthogonal transformations are considered (see Section 4.7). When X has rank strictly less than p, then none of the estimators are scale invariant. There is no unique unbiased least-squares estimator. Minimum length least-squares (MLLS) is obtained as a special case of all three regularization techniques. This estimator corresponds to the Moore–Penrose generalized inverse of X^TX, that least-squares estimator which minimizes the squared length $\hat{\beta}^T \hat{\beta}$. The rank of X will be less than p if the number of observations is no more than p but also possibly when by chance or construction columns of X are linearly dependent. The lack of invariance highlights the need for careful thought when considering the relative scales of variables for inclusion in regression. It would seem that scaling is used to make the regression coefficients equal in size. Since often little is known about the coefficients prior to data collection, scaling is sometimes applied routinely to take some cognizance of arbitrary units of measurement, and often avoided when units of measurement are all the same.

In using regularized estimators bias is traded for variance. Ordinary least-squares has no bias but may have a large variance. Within a quadratic loss framework the risk function usually considered is

$$\mathcal{R}(\hat{\beta}_R, \beta) = E\{(\hat{\beta}_R - \beta)^T L(\hat{\beta}_R - \beta)\}, \qquad (4.5)$$

where $L = I_p$ and $L = X^T X$ are popular choices, corresponding to estimation mean-squared error and prediction mean-squared error (at the model observations), respectively. Other values of L such as $X_2^T X_2$ for some $m \times p$ matrix X_2 would correspond to prediction at m points other than those of the model data; see for example Goldstein and Brown (1978). The description *prediction* follows immediately from considering the prediction of an entirely new set of n observations, Z, at the same model matrix, X, obeying the model

$$Z = \mu 1_n + X\beta + \epsilon$$

with

$$
\begin{aligned}
E(\hat{Z}_R - Z)^T(\hat{Z}_R - Z) &= E[n(\bar{Y} - \mu)^2 + \epsilon^T \epsilon \\
&\quad + \{X(\hat{\beta}_R - \beta)\}^T \{X(\hat{\beta}_R - \beta)\}] \\
&= (n+1)\sigma^2 + E\{(\hat{\beta}_R - \beta)^T X^T X(\hat{\beta}_R - \beta)\}.
\end{aligned}
$$

Prediction at the model points is much easier to accomplish than either estimation or prediction in directions with small eigenvalues.

An estimator $\hat{\beta}_R$ is said to *dominate* another estimator, say least-squares $\hat{\beta}$, with respect to a particular loss function, for example as in eqn (4.5), if the expected loss is everywhere in the parameter space at least as small as for $\hat{\beta}$, and strictly smaller for some β. An estimator which is dominated by another is termed *inadmissible*. Perhaps these words suggest stronger taboos than they ought. Inadmissible estimators are only inadmissible with respect to the particular chosen loss function and even then perhaps their expected loss might differ little over a relevant range of the parameter space. A more valid comparison arises from considering *a priori* likely values of the parameter space and a Bayesian analysis. Broadly all proper Bayes estimators (estimators that minimize the posterior expected loss when the prior distribution is proper, that is, integrates to 1) are admissible whatever the particular loss function chosen. Limiting Bayes estimators (such as when hyperparameters of the prior have limiting values which make the prior distribution unnormalizable, for example a prior density which is constant over the whole real line) may or may not be admissible. Regularization estimators such as ridge regression gain some justification from being closely allied to a Bayes estimator. The prior assumptions of this Bayes estimator allow one to gain insight as to when the estimator is on its home ground and will fare well.

4.2   Validation and cross-validation

The dangers of overfitting to a single set of data are widely recognized in statistical modelling. Methods such as least-squares and maximum likelihood will, untrammelled, fit the largest model. A technique that reluctantly overfits is desirable. Also it would be nice to be able to check that overfitting has not occurred. These two aspects are sometimes not kept as separate as they should be. You can attempt to validate a fitted model by collecting new data, untainted by the fitting process, and seeing how closely the fitted model predicts the new data. Preferably the model should fit a wide range of data-generating mechanisms, and it is important to map out the range of data for which the fitted model can be applied. However, the luxury of several sets of data may be absent in many day-to-day statistical analyses. Also what data one has may have relatively few observations. If you want to reserve some data for validation then an attractive economical way is provided by leave-one-out cross-validation. This considers

$$\sum (y_i - \hat{y}_{\bar{i}})^2, \tag{4.6}$$

where $\hat{y}_{\bar{i}}$ is the fitted value for the ith observation based on all y but the ith observation. This PREdiction Sum of Squares is sometimes known as the PRESS statistic (see Allen (1974)). If the number of observations, n, is large and the fit to $n-1$ observations is involved, then fitting n times to produce formula (4.6) can be computationally expensive. Often, however, the process can be speeded up by updating fits through subtraction of the ith observation from the fit with n observations (see Appendix D). If this is still too lengthy in computer time then you might choose a balanced or random subset of the data as the observations to omit, or divide the data into blocks of observations and leave out one block at a time.

Computational speed becomes more critical when the criterion eqn (4.6) is used to *choose* the fitted model. Cross-validatory choice as developed by Stone (1974) can offer a way of fitting of quite general applicability which to some extent inhibits overfitting. Amongst the class of models under consideration, the model which minimizes formula (4.6) is chosen. Stone (1977) has shown that this is asymptotically equivalent to the use of Akaike's criterion, discussed in Section 3.2.

Cross-validatory choice is not invariant to $n \times n$ orthogonal transformations of the vector of observations. In particular in the linear model (3.2), if Q is the matrix of eigenvectors of $\mathcal{X}\mathcal{X}^T$ then the transformed set of n observations Qy will have $(n-p-1)$ observations with mean zero and which do not involve the unknown parameters. The $(p+1)$ parameters are concentrated in just $(p+1)$ of the transformed observations. Such a situation is not promising for cross-validation. Intuitively it would seem that it would be better if the $(p+1)$ parameters were spread 'evenly' over the n observations: *generalized cross-validation* (GCV) does precisely this, reducing the new design matrix to a circulant matrix (see Golub *et al.* (1979)). What is more, when the estimator is linear in Y then the PRESS criterion leads to a computationally much simpler explicit minimization so that GCV can be quite generally preferred to cross-validation. For more discussion see the choice of ridge parameter in the next section.

Biased estimation methods are not invariant to changes of the relative scale of the explanatory variables. Some remarks about scaling have been made in Section 4.1. If automatic scaling, such as to equal variances, is part of the method then clearly the observations need to be rescaled every time an observation is removed in the leave-one-out process (see the corrigendum to Stone and Brooks (1990) for an instance of this). It is arguable whether this is necessary if really the scaling to common variance is just a rough and ready way of trying to achieve the impossible and equalize the true regression parameters. The true regression parameters cannot change because an observation is removed.

All commonly used biased regression estimators leave the intercept at its unshrunken value. Thus the estimated hyperplane goes through $(\bar{x}, \bar{y})$. If the x_{ij}, $j = 1, \ldots, p$ are centred variables over n observations, the $(n-1)$ subset $\{x_{lj},\ l \neq i\}$ is most conveniently re-centred as each observation $(i = 1, \ldots, n)$ is omitted. To do this centring, the mean of the $(n-1)$ subset $\bar{x}_{\bar{i}j}$ is given simply by

$$\bar{x}_{\bar{i}j} = -nx_{ij}/(n-1).$$

The new estimate of intercept is then the mean of the y_k, $k \neq i$. This has a correspondingly simple updating formula if the y_i were originally centred over the n observations. If despite our remark rescaling is desired, simple updating formulae may be devised similarly, see for example Stone and Brooks (1990).

Finally note that the use of cross-validation for model choice does not allow us to claim that we have validated the consequently chosen model. It has been chosen by sifting through all n observations. Further validation may well be desirable, perhaps by a nested further level of cross-validation: the 'two-deep' assessment of Mosteller and Tukey (1968, p. 147).

4.3   Ridge regression

The idea of adding a constant to the diagonal elements of the $X^T X$ matrix was used by several authors (for example, Marquardt (1963) and Hoerl (1962)) before the seminal papers of Hoerl and Kennard (1970a, 1970b) popularized the technique. The success of their promotion was due to theoretical and practical insights into the benefits, together with their use of the ridge trace graphic. Perhaps we now see that more was read into the existence theorem for improvement over least-squares than is strictly justified, but the technique is certainly valuable when applied with care. It is not a panacea and it does not solve the collinearity problem, but it can cope with it in a sensible way if used circumspectly.

The recipe as given by Hoerl and Kennard was as follows for model of eqn (3.2):

1. Centre the x-variables as in eqn (3.6), leading to model (3.7).
2. Scale the p x-variables so that the diagonal elements of $X^T X$ are n.
3. Plot the components of $\hat{\beta}_{RR}(k)$ versus k, where

$$\hat{\beta}_{RR}(k) = (X^T X + k I_p)^{-1} X^T y. \tag{4.7}$$

This plot is often called the ridge trace. The estimator of μ is traditionally taken to be $\bar{y}$ irrespective of k.

4. Plot the residual sum of squares as a function of k. In drawing this on the same graph as the ridge trace, a different (right-hand vertical axis) scale may be needed.
5. Choose a value of $\hat{k}$ of k which stabilizes the coefficient trace and at the same time does not penalize the sum of squares too much.

This last item clearly leaves much to the individual user's discretion. It also does not precisely define $\hat{k}$ as an estimator. The ridge trace has been criticized by a number of authors for its ease of misinterpretation (see for example Smith (1980)). It remains, however, at least a graphic teaching tool for demonstrating the effects of collinearity and the introduction of bias into an estimator. For routine application of ridge regression and estimation of the ridge constant, the ridge trace has been largely supplanted by well-defined estimators of k. In ill-conditioned near collinear problems the ridge trace may show dramatic swings of the coefficient estimates, with these even changing sign for relatively small changes in k. Figure 4.1 shows a ridge trace for the constructed data given in Table 4.1. These data are almost collinear with the first four variables adding to 10, apart from the first observation where they add to 11. The smallest eigenvalue of $X^T X$ scaled to correlation form is 0.001 and the largest 2.429, and the *condition number,* the ratio of largest to smallest, is rather large. After centring and scaling by the standard deviation of each of the six variables, ridge regression was applied to produce Fig. 4.1. The response was also centred and regression applied without an intercept term. Notice how the first four coefficient estimates change very rapidly for small k; the first coefficient even changes sign. The residual sample variance (dotted) increases slightly but steadily with k, confirming that least-squares ($k = 0$) does indeed minimize the error sum of squares. For a modest k the coefficients seem to have stabilized without incurring much increase in error variance. The broken line mean-squared error (MSE) curve confirms this; it would not normally be available as it shows the average squared difference between

Table 4.1. Design matrix for 12 observations on 6 variables and response

X_1	X_2	X_3	X_4	X_5	X_6	Y
8	1	1	1	0.541	-0.099	10.006
8	1	1	0	0.130	0.070	9.737
8	1	1	0	2.116	0.115	15.087
0	0	9	1	-2.397	0.252	8.422
0	0	9	1	-0.046	0.017	8.625
0	0	9	1	0.365	1.504	16.289
2	7	0	1	1.996	-0.865	5.958
2	7	0	1	0.228	-0.055	9.313
2	7	0	1	1.380	0.502	12.960
0	0	0	10	-0.798	-0.399	5.541
0	0	0	10	0.257	0.101	8.756
0	0	0	10	0.440	0.432	10.937

estimated coefficient and *true* coefficient, as used to simulate the response (scaled x-variables with coefficients: $\beta =$(0.603, 0.301, 0.060, -0.603, 0.904, 3.01) and $\sigma^2 = 1$).

It may be noted that centring of x-variables implies that the ridge hyperplane goes through the centroid $(\bar{x}, \bar{y})$ as with ordinary least-squares. The scaling in step 2 is a slight variation on the traditional recommendation to scale so that $X^T X$ is in 'correlation' form, for then the diagonal elements would be 1 not n. Here (2) is preferred because of its naturalness as the number of observations increase: essentially the meaning of β remains stable. Both scalings put the same relative weights on the x-variables, so that the two scalings are equivalent. The size of k may only be interpreted relative to the scaling (or lack of scaling) chosen. Our scaling by the standard deviation of the x-variables produces a k which is n times that from 'scaling to correlation form'.

The method as delineated by Hoerl and Kennard was accompanied by an existence theorem for the beneficial effect of eqn (4.7) for some $k > 0$ with respect to estimation accuracy (4.5) with $L = I_p$. Insight into this result and developments of the method are much more transparent after an orthogonal transformation of the model. Working with the model (3.7) with X centred, the rank of X is r and this is at most

$$\min(n - 1, p).$$

There exist matrices $T(n \times r)$ and $V(p \times r)$ with orthonormal columns such that

$$U = T^T Y = T^T 1_n \mu + T^T X V V^T \beta + T^T \epsilon,$$

with $V^T X^T X V = \text{Diag}(\lambda_1, \ldots, \lambda_r)$ and T consists of r eigenvectors of

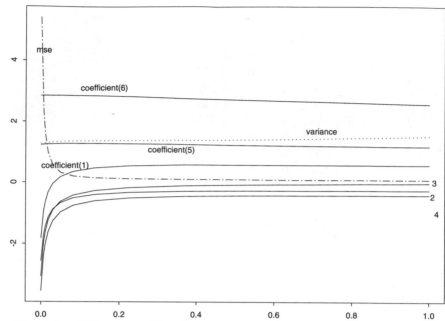

Fig. 4.1. Ridge trace of six ridge regression coefficients versus k, with variance and mean-squared error (from true simulated coefficients)

XX^T each orthogonal to the unit vector, so that the first term on the right is zero. The omitted column-vector of T is proportional to the unit vector and provides that linear combination of the Ys that has mean the intercept term. Here

$$T^T X V = \mathrm{Diag}(\sqrt{\lambda_1}, \ldots, \sqrt{\lambda_r})$$

is the singular value decomposition of X (see Appendix D). The $\lambda_1 \geq \ldots \lambda_r > 0$ are the r non-zero eigenvalues of $X^T X$. The number of non-zero eigenvalues, r, is the rank of X. Because of the orthogonality , the error structure and estimation are unaffected by this transformation to canonical form. This is also true of the loss function in eqn (4.5) with $L = VV^T$ or I_p, provided X has rank $r = p$. Letting

$$V^T \beta = \alpha,$$

the model becomes

$$
\begin{aligned}
U_i &= \sqrt{\lambda_i}\alpha_i + \epsilon_i, \ i = 1, \ldots, r & (4.8)\\
U_i &= \epsilon_i, \ \ i = r+1, \ldots, n-1. & (4.9)
\end{aligned}
$$

The ridge estimator is

$$\hat{\alpha}_{iRR} = U_i\sqrt{\lambda_i}/(\lambda_i + k)$$

for $i = 1, \ldots, p$, actually zero for zero eigenvalues with $r < i \leq p$. Otherwise, for non-zero eigenvalues, $i = 1, \ldots, r$ the least-squares estimator of α_i, denoted simply $\hat{\alpha}_i$, is $U_i/\sqrt{\lambda_i}$ and

$$\hat{\alpha}_{iRR} = \{\lambda_i/(\lambda_i + k)\}\hat{\alpha}_i. \tag{4.10}$$

For α-coefficients corresponding to zero eigenvalues the least-squares estimator is indeterminate, but is zero for the minimum length least-squares estimator. Thus for $k > 0$, $\hat{\alpha}_{iRR} < \hat{\alpha}_i, i = 1, \ldots, r$ so that $\hat{\alpha}_{RR}$ is $\hat{\alpha}$ shrunken towards zero, and

$$\hat{\alpha}_{iRR} = w_i 0 + (1 - w_i)\hat{\alpha}_i, \tag{4.11}$$

a weighted average of zero and the least-squares estimator with weight

$$w_i = k/(\lambda_i + k).$$

Thus zero has a special feature as the point of attraction. The effect of a particular k is greater for components with small λ_i. Note, however, that with respect to the original non-canonical model, eqn (3.7), individual coefficients estimators $\hat{\beta}_{iRR}$ may not be shrunken versions of $\hat{\beta}_i$ (that is, $|\hat{\beta}_{jRR}| > |\hat{\beta}_j|$ for some j); only the overall length of the p-vector $\hat{\beta}_{RR}$ has length less than $\hat{\beta}$. It was the excessive length of least-squares estimators following from the expectation

$$E(\hat{\alpha}^T\hat{\alpha}) = \alpha^T\alpha + \sigma^2 \sum(1/\lambda_i),$$

and greatly inflated by small λ_i, that attracted statisticians to shrinkage estimators and more particularly ridge regression. The locus of $\hat{\beta}_{RR}$ as k increases from zero, for least-squares, to k infinite, when the estimator is the zero p-vector, is as follows. The least-squares residual sum of squares as a function of β in p-space is elliptical with principal axes proportional to $\sqrt{\lambda_i}$ and centred on the least-squares point. If the rank of X is r less than p, then the least-squares function does not change in a $(p-r)$ dimensional hyperplane, and is elliptical in the linear r-space orthogonal to this. If you take concentric p-spheres centred at the origin, then the points at which spheres touch the nested ellipses is the locus of $\hat{\beta}_{RR}$ as the radius of the spheres vary. This is shown in two dimensions in Fig. 4.2. The locus begins at $k = 0$, the maximum likelihood estimate (the centre of the ellipses), and spirals in to the origin at $k = \infty$. At this origin the locus is locally the normal to the ellipse through the origin, and all ridge estimates lie in the sector between this normal and the line from the origin through the centre

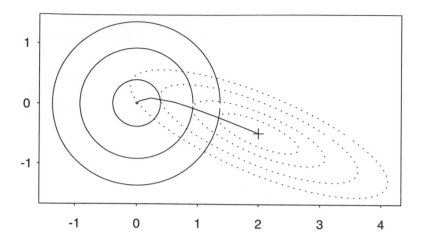

Fig. 4.2. Likelihood contours (ellipses) and touching circles: locus of the ridge estimator with k is curve between centres of ellipse ($k = 0$) and circle ($k = \infty$)

of the ellipse (MLE). If the situation is near collinear in one dimension then in that dimension the ellipse will be very narrow, with a small eigenvalue. In the collinear limit it is the point on the line nearest to the origin. Indeed in the singular case the limit as $k \to 0$ provides the unique minimum length least-squares estimator.

This geometry is an interpretation of the following constrained minimization problem. Let b be any estimate of β. Suppose you take b to be on a contour of the profile likelihood of β,

$$(b - \hat{\beta})^T X^T X (b - \hat{\beta}) = \text{const} \qquad (4.12)$$

and you wish to minimize the squared length $b^T b$ subject to this constraint. This is the complementary problem to the constrained minimization of eqn (4.1). In the canonical form with $V^T b = a$, minimize $a^T a$ subject to $\sum (a_i - \hat{\alpha}_i)^2 \lambda_i = \text{const}$. With the Lagrange multiplier $(1/k)$, the Lagrangian is

$$\sum a_i^2 + (1/k) \sum (a_i - \hat{\alpha}_i)^2 \lambda_i.$$

Differentiating partially with respect to a_i and setting the result equal to zero, gives

$$a_i = \lambda_i \hat{\alpha}_i / (\lambda_i + k),$$

the ridge estimator. Also

$$\sum a_i^2 = \sum \hat{\alpha}_i^2 \lambda_i^2 / (\lambda_i + k)^2 = f(k)$$

where $f(k)$ is monotone in k and $f(0) = \sum_1^r \hat{\alpha}_i^2$, $f(\infty) = 0$.

One further motivation for the estimator is with respect to a simple Bayesian prior distribution on β, or, equivalently on α in the canonical model. Suppose β_i are *a priori* independent normal with mean zero and variance σ_β^2, with σ_β^2 assumed known. Furthermore, the intercept term μ is taken to have a vague prior distribution, uniform over the real line. Then *a posteriori* μ is normal with mean $\bar{y}$ and variance σ^2/n. Independently of this, with centred y-data, the posterior density of β is proportional to the likelihood times the prior, that is, proportional to

$$\exp(-1/2)\{(y - X\beta)^T (y - X\beta)/\sigma^2 + \beta^T \beta/\sigma_\beta^2\}.$$

The exponential argument is quadratic in β and thus β is normal *a posteriori*. Completing the square in the quadratic form in β this multivariate normal posterior has mean $\hat{\beta}_{RR}$ and covariance matrix

$$\sigma^2 (X^T X + kI_p)^{-1} \tag{4.13}$$

where $k = \sigma^2/\sigma_\beta^2$. If σ_β^2 is very large then the prior variances of the β_i are large and k is near zero, whence the posterior mean, $\hat{\beta}_{RR}$, approaches the least-squares solution, or MLLS when X has rank less than p. For $k > 0$ the posterior covariance matrix eqn (4.13) is always of full rank p even when X has rank less than p. Note also that the implied prior for $\alpha = V^T \beta$ is identical to that for β because of the orthogonality of V and equality of prior variances of β_i.

The implication of this Bayesian motivation is that the ridge estimator will tend to do well if these prior assumptions are met. Broadly the regression coefficients should be clustered around zero and look like a random sample from a zero mean normal distribution. The variables might be viewed as speculative. This would not be the case if, for example, there were just a few coefficients large in absolute size and the rest small. You may sometimes be confident of symmetry or exchangeability (see de Finetti (1974)) by substantive prior knowledge formed of past experience and appropriate scaling of the x-variables. Some asymmetry may also be present and you might also wish to identify non-speculative variables of known pedigree and treat these separately with a much smaller or zero ridge constant.

It is clear that the procedure is not, and should not be, scale invariant. If $x_j \to x_j/c$ then $\beta_j \to c\beta_j$ and the prior variance of β_j changes to $c^2 \sigma_\beta^2$. Therefore rather than automatically scale the variables to a common variance it would be better if possible to effect a scaling c_j for x_j such that the corresponding regression coefficients are similar in absolute size.

A further instructive underpinning of the ridge estimator is as a convex combination of the least-squares estimators for the 2^p different submodels formed from whether or not each of the p variables is or is not included in the model. The weights are a sensible balance of prior and sample information (see Leamer and Chamberlain (1976)).

4.4 Ridge properties and estimators

Many properties follow rather simply from the canonical form (eqns (4.8, 4.9)) and the relationship of the ridge estimator to the least-squares estimator in eqn (4.10), at least when k is regarded as known. Deeper and harder results arise when k is estimated by $\hat{k}$, some explicit function of the data. In this section the less realistic fixed k results are first given. Then estimators $\hat{k}$ of k are introduced and some properties of $\hat{\beta}_{RR}(\hat{k})$ using Stein's technique of unbiased risk estimation are presented with further details in Appendix C.

Firstly the existence theorem of Hoerl and Kennard is proved. Estimator (4.10) has bias, in canonical form,

$$E(\hat{\alpha}_{iRR} - \alpha_i) = \lambda_i \alpha_i / (\lambda_i + k) - \alpha_i = -k\alpha_i / (\lambda_i + k). \qquad (4.14)$$

From the same equation (4.10) the variance of $\hat{\alpha}_{iRR}$ is

$$\{\lambda_i / (\lambda_i + k)\}^2 \sigma^2 / \lambda_i = \lambda_i \sigma^2 / (\lambda_i + k)^2. \qquad (4.15)$$

Both the bias and variance formulae also hold with singularities corresponding to $\lambda_i = 0$. The mean-squared error is the variance plus the square of the bias and consequently, in weighted form,

$$\begin{aligned}
E \sum_1^p L_i (\hat{\alpha}_{iRR} - \alpha_i)^2 &= \sigma^2 \sum L_i \lambda_i / (\lambda_i + k)^2 \\
&\quad + k^2 \sum L_i \alpha_i^2 / (\lambda_i + k)^2 \\
&= \mathcal{V}(k) + \mathcal{B}(k) \qquad (4.16)
\end{aligned}$$

For estimation mean-squared error you would set $L_i = 1$, whereas for prediction mean-squared error (at the model matrix) $L_i = \lambda_i$. The first term of eqn (4.16) identified as $\mathcal{V}(k)$ is such that

$$d\mathcal{V}(k)/dk|_{k=0} < 0$$

whereas for the bias squared term

$$d\mathcal{B}(k)/dk|_{k=0} = 0.$$

The variance term has a negative slope at $k = 0$ whereas the slope of the bias term is zero. Thus moving from $k = 0$ to $k > 0$ decreases the mean-squared error by decreasing the variance without any increase in bias. This is in essence Hoerl and Kennard's existence theorem. Unfortunately such an argument does not tell you which value of $k > 0$ to choose. It claims only the existence of a $k > 0$, however small. Fortunately in practice and in circumstances already indicated by Bayesian arguments often the improvement over least-squares may hold for $k < k_0$ when k_0 is sizeable and the mean-squared error does plummet from $k = 0$ for this range of k.

The existence result could be proved for shrinkage to any fixed point. The zero shrinkage point in eqn (4.11) is not special. This further undermines the value of the existence result. Also it would be nice to think that the ridge regression estimator dominates the least-squares estimator. This cannot be, for the simplest of reasons as pointed out by Thisted (1976): eqn (4.7) is not an estimator unless a value of k is specified. It forms a class of estimators as k is allowed to change. Sometimes a generalization of ridge regression is suggested in which a different ridge constant is applied to each variable. This greater latitude in the class of estimators enables even more grandiose claims to be made about the efficiency of the method, and suffers from the unclearly addressed need to estimate many hyperparameters.

To properly investigate ridge regression as an estimator it is natural to look at some data-dependent choices for $\hat{k}$. Many possible estimators $\hat{k}$ of k have been suggested. The ridge trace method is discounted as it does not give an unambiguous estimator. Here five possibilities are listed. In these definitions hat (ˆ), over the parameters (α, β, σ^2) of the regression model denotes the least-squares estimator. They are defined in a notation allowing singular X of rank $r < p$, when for least-squares interpret minimum length least-squares. When there are more variables than observations and $r = (n - 1)$, an estimate of the residual variance must be supplied from other sources.

1. Cross validation estimate, $\hat{k}_{CV}$, for appropriately scaled data.
   For a discussion of scaling see the introduction to this chapter, and for scaling in cross-validation see Section 4.2. A version, $\hat{k}_{GCV}$, which spreads the parameters evenly over the n observations, developed by Golub *et al.* (1979), is called generalized cross-validation and discussed in Section 4.2. They choose k to minimize

$$||I - A(k)y||^2/[\text{trace}(I - A(k))]^2$$

   where $A(k) = X(X^T X + kI)^{-1}X^T$. Notice that this optimization problem is completely vectorized and does not involve looping inherent in leave-one-out algorithms.

2. As Hoerl *et al.* (1975), but modified to conform with Stein's estimator in the orthogonal case:

$$\hat{k}_{MHKB} = (r-2)\hat{\sigma}^2/\hat{\beta}^T\hat{\beta}.$$

This and the next estimator of k are motivated by the temporary adoption of a Bayesian analysis, with $k = \sigma^2/\sigma_\beta^2$ and rather simple forms of empirically estimating σ_β^2.

3. As Lawless and Wang (1976), but modified again to conform with the Stein estimator in the orthogonal case:

$$
\begin{aligned}
\hat{k}_{MLW} &= \{(r-2)\hat{\sigma}^2\text{trace}(X^TX)\}/(r\hat{\beta}^T X^TX\hat{\beta}) \\
&= \{(r-2)\hat{\sigma}^2\text{trace}(X^TX)\}/(r\hat{y}^T\hat{y}).
\end{aligned}
$$

Here y as well as x-variables are centred over the n observations, so that $\hat{y}$ requires no intercept adjustment.

4. Minimum unbiased risk estimator; $\hat{k}_{MUR}$ minimizes an unbiased estimator of the mean-squared error or risk. For fixed k an unbiased estimator of the risk is easiest to derive from the canonical form. From eqns (4.14), (4.15), and (4.16), with $L_i = 1$ (zero for inestimable α_i), an unbiased estimator of the risk is

$$\hat{\sigma}^2\sum(\lambda_i - k)/\{\lambda_i(\lambda_i + k)\} + k^2\sum\hat{\alpha}_i^2/(\lambda_i + k)^2.$$

5. Maximum integrated likelihood. This is an approximate Bayesian method pioneered as 'type II' maximum likelihood by I. J. Good, see for example Lindley and Smith (1972) and empirical Bayes procedures. Using, say, an independent normal with mean zero and variance σ_β^2 for β, the marginal distribution of the data given the parameter σ^2 and hyperparameter σ_β^2 can be formed. Type II maximum likelihood maximizes the likelihood of the two parameters from this likelihood derived from the marginalized distribution. The value $\hat{k}_{MML}$ is given by the ratio of marginal maximum likelihood estimators of σ^2 and σ_β^2. This method is applied by Anderssen and Bloomfield (1974), who just call it maximum likelihood. Nowadays with better numerical integration tools (for example Tierney and Kadane (1986)) a more fully Bayesian analysis is straightforward. A Bayesian statistician may not wish to be constrained to the ridge class, because of its implicit normal prior for β and also because it would be desirable to average over the posterior distribution of k given the data rather than use a simple plug-in estimator for k.

Various simulation studies (for example, Hoerl $et\ al.$ (1975), and Lawless and Wang (1976))) offer only partial guidance, especially as their simulations are somewhat favourable to their methods. As a critique of ridge regression and a lesson in how to set up a simulation study efficiently, Thisted (1976) is excellent. He does not consider the cross-validation estimator, $\hat{k}_{CV}$, which would be an exception to the following points:

1. for a particular design the ridge estimator is a function of the p sufficient statistics for β and the one for σ^2;

2. all simulations may be performed in the canonical form of the model.

Taken together these imply that a single simulation requires only $(p+1)$ random variables, p univariate normal and one chi-squared.

No simulations are going to be entirely convincing, and ridge regression stands on a variety of successful applications, and sometimes falls on unsuccessful ones. A version of ridge regression has been successfully used by BBC television for election-night forecasting on five separate British General Elections from 1974 until 1987; see Brown and Payne (1975), Payne and Brown (1981) for forecasting the British Election to the European Parliament.

A few remarks on the estimators of the ridge constant k listed above may be appropriate. Cross-validation provides a worthy estimator, but can be computationally rather intensive when the sample size is large. It does not, however, require a separate estimate of σ^2 and can be used even when $n < p$ so that the linear relationship is underdetermined and least-squares would produce a 'perfect' fit. The generalized cross-validation estimator is particularly attractive computationally for users of interactive vectorized languages such as S and MATLAB. The estimator $\hat{k}_{MHKB}$ may be too small in near collinear situations when the eigenvectors corresponding to small eigenvalues are uncorrelated with y. This insight is based on the fact that the least-squares divisor $\hat{\beta}^T\hat{\beta}$ may then be rather inflated. The ridge constant estimator $\hat{k}_{MLW}$ may be preferable in these circumstances as it downweights $\hat{\alpha}_i^2$ by the small λ_i eigenvalue. The estimator $\hat{k}_{MUR}$ is somewhat untried in the literature but is very much in the spirit of Mallows (1973) discussed in Section 3.2.

Computational aspects of simulation studies were mentioned above, in which the usefulness of the singular value decomposition was emphasized; see Appendix D. The singular value decomposition is a basic tool of many good statistical packages and environments, for example S, LISP-STAT, and GENSTAT. Ridge regression programs may be easily developed from the least-squares $\hat{\alpha}$ of the canonical model and then converted back using

$$\hat{\beta}_{RR} = V\hat{\alpha}_{RR}.$$

This simple computational route was used to fit a linear model by ridge regression for the second component of the detergent data, Example (1.5). The data consist of twelve observations on 1168 absorbance variables. It is debatable whether such co-measurable variables should be scaled to have the same variance. Here we decided against any relative scaling, and just scaled all variables by the median of the 1168 standard deviations. Taking

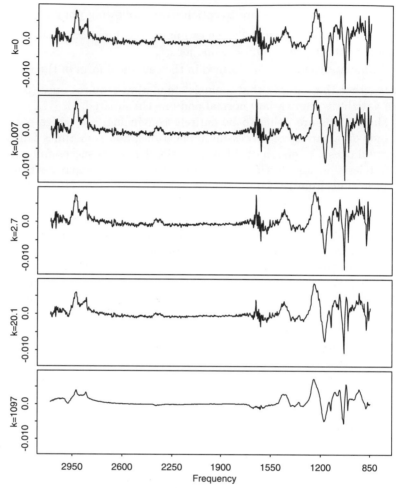

Fig. 4.3. Detergent data, ridge regression coefficients for various ridge constants (k) by frequency (cm^{-1})

an equally spaced grid of k values on the logarithmic scale gave a minimum cross-validated root mean-squared prediction error of 0.253 at $k = 20$. The ridge estimated regression coefficients for a range of values of k are plotted in Fig. 4.3. Here $k = 0$ is minimum length least-squares. Notice how excessive fuzziness is reduced as k is increased, mostly so by k as large as this optimal value of 20.

One alternative computational route of at least pedagogical interest arises from viewing the ridge estimator as arising from data augmentation. Working with centred y and as well as x-variables, augment the $n \times p$ matrix X with p further rows of the form $\sqrt{k}I_p$, calling the resultant $(n + p) \times p$

matrix X_k, Now augment the n-vector y with p zeros, forming the $(n+p)$-vector y_k. Applying the least-squares formula to this new problem with $(n+p)$ data points gives the ridge estimator, since

$$(X_k^T X_k)^{-1} X_k y_k = (X^T X + kI_p)^{-1} X^T y.$$

This was noted by Marquardt (1970). This also opens up the possibility of performing a QR-decomposition of the matrix X_k, avoiding the numerically undesirable squaring of X in the 'normal' equations. One possible pitfall of the augmentation approach is that a general set of least-squares algorithms will output various quantities connected with the augmented data and many of these may be inappropriate. For example, the residual sum of squares will be valid only for fitting the augmented data including the p zero values of response.

Sufficient conditions for the dominance over least-squares of a class of adaptive ridge estimators are given in Thisted (1976) and generalized to multivariate ridge regression in Brown and Zidek (1980, 1982). The methodology also gives an unbiased estimator of the improvement in risk. The derivations are rather technical and revolve around the unbiased risk estimation technique of Stein (1973). A simplified version is given in Appendix C. The full rank case is considered. The type of result that emerges is, for example, that the modified Hoerl, Kennard and Baldwin estimator dominates least-squares with respect to estimation quadratic loss if as given by eqn (C.8),

$$(h_p)^2(p+2) < 2p\overline{(h^2)}$$

where $h_i = \lambda_i^{-1}$, $\overline{(h^2)} = \sum h_i^2/p$ and λ_p is the smallest element of $\{\lambda_1 \geq \cdots \geq \lambda_p\}$. This is a stringent requirement and effectively demands that the eigenvalues be not too disparate. Such conditions are not necessary, but the development of the result shows that one has little hope of dominating least-squares by ridge-style estimators when the condition number, λ_1/λ_p, is large, this being the case of near collinearity.

Stein estimation (1961) shrinks by a constant amount, giving $\hat{\alpha}_{iS} = c\hat{\alpha}_i$ for $c = 1 - (p-2)/\|\alpha\|^2$. When the eigenvalues are all equal, the design consists of orthonormal columns, and both ridge estimators with k estimated by $\hat{k}_{MHKB}$ and $\hat{k}_{MLW}$ reduce to the Stein estimator. For the Stein estimator in these circumstances dominance over least-squares is assured. When the eigenvalues are unequal but interest turns to prediction loss at the configuration of model points ($L = X^T X$), Stein-type estimation can still guarantee dominance over least-squares where ridge cannot, although the condition above becomes less stringent (see Appendix C). This dominance property of Stein-type estimation is further exploited by Copas (1983). The argument for ridge estimation over Stein estimation does not rely on dominance over least-squares; it is born more of the nature of the shrinkage and for many applications its more natural implied Bayesian prior assumptions than those implied by Stein estimation in the unequal eigenvalue case.

Finally, much has been said about ridge estimation, but little has been said about ridge inference, confidence intervals for β, and prediction intervals for future Y. The literature is mostly concerned with estimation. The classical approach to inference is to apply standard sampling intervals ignoring the shrinkage aspect of estimation (see Obenchain (1977)). On the other hand, if you are comfortable with the Bayesian underpinning and its prior assumptions, then you might work with covariance matrix eqn (4.13), and as a first approximation plug in an estimate of k. For a future response, Z, at an observed p-vector x on the same scales as the observations making up X, the normal Bayesian predictive distribution has mean $\bar{y} + x^T \hat{\beta}_{RR}$ and variance

$$\sigma^2 \{ 1 + 1/n + x^T \mathcal{G} x \} \tag{4.17}$$

with $\mathcal{G} = (X^T X + kI_p)^{-1}$.

4.5   Principal components regression

Principal components regression was previewed in the introduction to this chapter. Coefficient vector β is estimated by estimator (4.3) in conjunction with an 'inverse' of $X^T X$ given by eqn (4.4). Often the label-set S_ω of eigenvalues of $X^T X$ chosen corresponds to the ω largest eigenvalues. The p orthogonal eigenvectors v_j, $j = 1, \dots, p$ of $X^T X$ provide p new canonical variables, the *principal components*, as linear combinations, $v_j^T x$ of the original variables, with jth sample variance λ_j. Sometimes principal components form a set of plausible constructed variables with interpretable meanings. For example, given the logged measurements of height, breadth, and length, then if $v_1 = (1/\sqrt{3}, 1/\sqrt{3}, 1/\sqrt{3})$, the first principal component is the sum of these logged dimensions and is a suitable measure of logged volume. On prior grounds you might wish to focus purely on this first principal component.

As a more substantial illustration, in the twelve-observation detergent data of Example 1.5, X consists of a 12×1168 matrix of centred absorbances. This allows up to $r = \min(11, 1168) = 11$ principal components. Fitting the ω principal components so as to correspond to the largest singular values of X is easily achieved using the singular value decomposition, as in the canonical reduction of the linear model leading to eqn (4.8). Here the principal components regression estimate of α for ω principal components is the r-vector

$$\hat{\alpha}_\omega = (U_1/\sqrt{\lambda_1}, \dots, U_\omega/\sqrt{\lambda_\omega}, 0, \dots, 0)$$

and that for the p-vector β is given as

$$\hat{\beta}_\omega = V\hat{\alpha}_\omega,$$

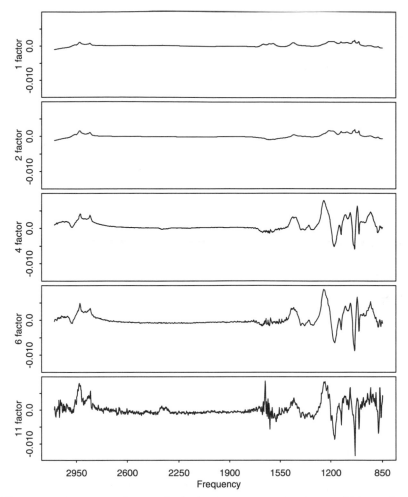

Fig. 4.4. Detergent data, principal component regression coefficients for various numbers of components fitted, by frequency (cm^{-1})

where the 1168×11 orthonormal matrix V is from the singular value decomposition. These estimated regression coefficients are plotted for selected values of ω in Fig. 4.4. With all 11 components fitted the estimator is minimum length least-squares. Fitting fewer components might seem desirable if no more than to avoid fuzziness introduced by noise.

It may be necessary to rely on an automated procedure for choosing principal components. If the sum of principal component variances, the sum of eigenvalues (or trace) of $X^T X$, is called the *total* variance, then ω might be chosen so that the proportion of total variance of x-variables utilized is suitably high, perhaps at least 90 per cent. Other rules based on choosing only sizeable eigenvalues are common. Such choices have the inherent weakness of ignoring information which has not been verified as unimportant. The directions of admittedly small variance may correlate well with the response. Of course if an eigenvalue λ_j is close to zero then the corresponding canonical coefficient α_j will have high variance (σ^2/λ_j) of estimation. It is often observed that such small eigenvalues represent essentially noise and can be ignored with impunity. Be this as it may, the policy of choosing the label set on the basis of eigenvalues of $X^T X$ without recourse to the response is bound to come unstuck sometimes. Better is the hybrid strategy (b) alluded to in Section 4.1. The label set of chosen eigenvalues and eigenvectors is based on the correlation of principal components with Y. Cross-validatory prediction squared error might give the number and identity of those eigenvectors. Provided the sample size is small then the corresponding computational burden will not be excessive. In envisaging this burden you should remember that new principal components need to be calculated for each observation left out in turn. For a particular partition of the data the principal components are uncorrelated amongst themselves and least-squares estimators of α_j are correspondingly uncorrelated. The principal components may be ordered in importance by the size of their correlation with the response. One computational short cut would approximate the full cross-validatory choice by taking the order of introduction of principal components as fixed by the full n-sample data. The number, ω, of principal components would remain to be chosen. Such a procedure might not be sensible if principal components changed order with one another when a point was omitted.

A less computationally intensive choice of number of components might be adapted from Mallows C_p (1973) and the earlier discussion of Section 3.2. For a fixed set of chosen principal components, the principal components regression estimator is linear in Y and an unbiased estimator of the risk is readily specified, and hence a graphical choice of ω may be devised. Other strategies are described in Jolliffe (1986). One particularly simple strategy he suggests is to order the principal components from largest to smallest eigenvalue and work backwards from the smallest testing for statistically significant regression coefficients. The first significant coefficient corresponding to λ_ω would determine the largest ω principal components to be used in regression. This is viewed as a compromise strategy taking account of both x-variance and correlation with y. Broadly, it hurts if you omit a principal component which has a significant correlation with the response; principal components corresponding to large eigenvalues are relatively harmless to regression whether or not they are correlated with y.

Under normality, sample principal components and eigenvalues are maximum likelihood estimators of their population counterparts, and asymptotic distribution theory is available (see for example Kshirsagar (1972)). If in regression you then depict certain principal components as having no or insignificant correlation with the response Y, estimates and predictions are again maximum likelihood conditional on the truth of no relationship. Since such hypotheses are tested by recourse to data their truth is not certain and estimation and prediction intervals will ignore this source of uncertainty and be overconfident. There is a parallel here with ignoring uncertainty in using plug-in estimators for the constant in ridge regression. The crudest prediction interval for a future response at an observed explanatory p-vector x may be obtained from eqn (4.17), the variance of the normal Bayesian predictive distribution, with $\mathcal{G}$ given by eqn (4.4).

4.6 Partial least-squares regression

Partial least-squares regression (PLSR) arose out of the systems analysis approach to prediction modelling developed by Herman Wold and his co-workers in the 1960s. It owes much to Svante Wold and Harald Martens for its development (see Wold *et al.* (1983)). More recently, for statisticians, the method has become clearer and stripped of its algorithmic emphasis in papers by Helland (1988) and Stone and Brooks (1990). The maximization of covariance to define PLSR was earlier described by Frank (1987). Stone and Brooks link together PLSR, PCR, and ordinary least-squares (OLS) as particular techniques along a continuum of possibilities. We adopt their two-stage approach to define PLSR. More discussion of Stone and Brooks's *continuum regression* appears in Section 4.8. PLSR aims to relate two sets of variables. Most often one wishes to predict a single response variable Y by means of p variables $x_1, x_2, ..., x_p$, as in model (3.7) with the x-variables centred. In regression one wishes to form a linear combination of the p explanatory variables to predict Y. The first stage constructs a few orthogonal latent variables by maximizing *covariance*. In the second stage you apply ordinary least-squares to the regression of y on these latent variables.

Explicitly consider that set of p constants $c_1, ..., c_p$, which are standardized to have length unity, that is $\sum c_j^2 = 1$, and which maximize the sample covariance, $\sum_1^n y_i t_i$, with $t_i = \sum_1^p x_{ij} c_j$. This leads to the first *latent factor*, say the vector, $t^{(1)}$, that is, the set of n entries of t corresponding to the maximizing set of p c-values. One then sequentially generates further latent factors, $t^{(2)}, ..., t^{(\omega)}$, where ω is carefully chosen (see later cross-validatory choice). The n entries of $t^{(2)}$ are constructed from that set of constants c which maximize the sample covariance with y with the constraint that $t^{(2)}$ is orthogonal to $t^{(1)}$. In other words $\sum_1^n t_i^{(1)} t_i^{(2)} = 0$.

Formally and generally if

$$d = X^T y \text{ and } D = X^T X \tag{4.18}$$

then it is first necessary to maximize $(c^T d)^2$ subject to the constraint $c^T c = ||c||^2 = 1$. Now by the Cauchy–Schwarz inequality,

$$(c^T d)^2 \leq ||c||^2 ||d||^2$$

with equality if and only if $c \propto d$ and hence

$$c_1 = d/||d||.$$

Secondly, maximizing $(c^T d)^2$ subject to both $||c|| = 1$ and D-norm orthogonality to c_1, that is $c^T D c_1 = 0$, may be achieved using Lagrange's method with two undetermined multipliers. Accordingly let

$$F = (c^T d)^2 - \nu_1(c^T c - 1) - 2\nu_2 c^T D d,$$

differentiating partially with respect to the vector c and setting the result equal to the zero vector gives

$$2(c^T d)d - 2\nu_1 c - 2\nu_2 D d = 0. \qquad (4.19)$$

Multiplying on the left by c^T gives

$$\nu_1 = (c^T d)^2, \qquad (4.20)$$

and now multiplying eqn (4.19) by $d^T D$ on the left one obtains

$$\nu_2 = (c^T d)(d^T D d/d^T D^2 d)$$

and this with eqn (4.20) may be plugged back into eqn (4.19) to give

$$c_2 \propto d - (d^T D d/d^T D^2 d)D d.$$

Continuing in this way you may derive the remaining canonical covariance variables, $c_3, \ldots, c_\omega$. In fact, as is suggested by the forms of c_1 and c_2, it may readily be seen that $c_1, \ldots, c_\omega$ are spanned by

$$\{d, Dd, \ldots, D^{\omega-1}d\} \qquad (4.21)$$

where these are linearly independent. In the numerical analysis literature, $d, Dd, \ldots$ is called a Krylov sequence. The dimension of the space spanned by this sequence will be the maximum number of factors that the PLSR algorithm can give. This maximal space is also spanned by what Helland (1988) calls the *relevant factors* of $D = X^T X$, that is the eigenvectors of D with non-zero components along $d = X^T y$, one for each distinct eigenvalue in case these should be degenerate. If d were an eigenvector of D then only one iteration will give the full least-squares solution, minimum length if D is singular.

Remark 1. The maximizing covariances will get progressively smaller as one continues sequentially generating latent factors in such a way that they are orthogonal to the previous latent factors. This has to be so since successively more side conditions are imposed on the maximization of covariance. After correcting for the mean there are $\min(n-1, p)$ degrees of freedom, after fitting this number of factors, the sample covariance must be identically zero for further factors. The actual maximum number of factors ω^* that may be fitted may be considerably less than $\min(n-1, p)$, see remarks about Krylov sequences above. Some PLSR software packages seem in their algorithm to fail to check whether the covariance is effectively zero. The covariances may be truly zero early in the sequential construction. The first covariances that are actually zero may not be exactly computed as zero because of rounding errors. This will not matter except when more than the upper bound of $\min(n-1, p)$ factors are allowed blithely to be fitted!

Remark 2. Whilst PLSR forms latent factors by a *canonical covariance* analysis of Y and X, PCR forms its latent factors by a canonical *variance analysis* of X alone. Furthermore, if you had focussed on the *correlation* between Y and X then this is just the standard least-squares method, and there would be *only one* latent factor. This in the context of more variables than observations would have $(p + 1 - n)$ degrees of indeterminism (see Sundberg and Brown (1989)). Standard canonical *correlation* analysis, for univariate or multivariate Y, focuses on correlation rather than covariance and is quite distinct from PLSR.

The second stage of the PLSR method comprises regressing y on the $\omega \leq \min(n-1, p)$ orthogonal latent factors, forming the predictor

$$\hat{Y} = t^{(1)}b_1 + ... + t^{(\omega)}b_\omega. \tag{4.22}$$

The combination of latent factors in eqn (4.22) can be written as a linear combination of the original x-variables, providing the PLSR estimator of β in the linear model of eqn (3.7).

4.6.1 DETERGENT DATA

As an illustration consider the detergent Example (1.5), with the response component 2. For comparison the same example has already been used to illustrate ridge and principal components regression. There are just $n = 12$ observations but $p = 1168$ frequencies. We assume the model (3.7). If you allow one degree of freedom for the estimate of the mean parameter then with just eleven factors you can predict the y data perfectly, with zero error in the fitted relationship. The estimates of β across the 1168 frequencies are given in Fig. 4.5 progressively with 1, 3, 4, 7, 11 factors fitted. The saturated 11-factor model gives that least-squares estimate

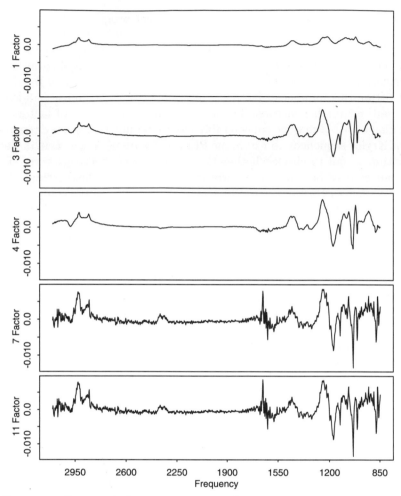

Fig. 4.5.   Detergent data, regression coefficients by wavelength estimated by partial least-squares with varying numbers of factors, by frequency (cm^{-1})

which has minimum length, the minimum length least-squares estimator (MLLS). This estimator can be obtained by other methods of estimation, such as least-squares applied to eqn (3.2) with estimator (4.3) where $\mathcal{G}$ is the Moore–Penrose generalized inverse of the 12×12 sum of products matrix $X^T X$. Another way of obtaining MLLS would be by principal component regression by including all factors (≤ 11) which have non-zero eigenvalues in the eigenanalysis of $X^T X$. For these data the 11 factor curve of coefficients in Fig. 4.4 is the MLLS set of coefficient estimates. A further way is to allow $k \to 0$ in ridge regression, see the first graph ($k = 0$) of Fig. 4.3.

Notice how in Fig. 4.5 the use of just one factor shows a very suppressed

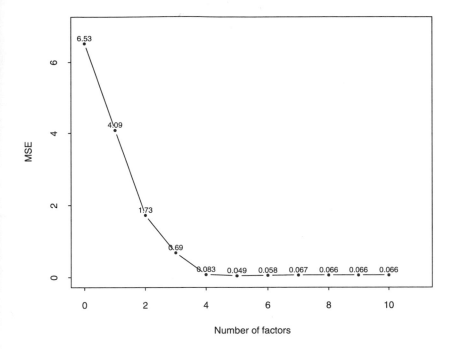

Fig. 4.6.   Detergent data, 1168 wavelengths, cross-validated mean-squared error by number of factors fitted by partial least-squares

set of estimates. By 3 factors the final shape has largely emerged, and perhaps shows its most satisfactory form at 4 factors. After this high frequency noise begins to appear. Indeed a plot of cross-validated prediction mean-squared error in Fig. 4.6 shows a minimum of 0.049 at 5 factors. Here with one observation omitted, 10 factors constitutes the saturated model as applied to subsets of 11 observations (mean corrected). The detergent actually contains five different ingredients, as listed in Table 1.4, adding up to 100 per cent, so that from prior considerations you would hope that four factors might be enough. This is not so for this component although it only requires one more factor. Other components need more factors. The continuity of the regression coefficient estimates in Fig. 4.5 is perhaps surprising in that partial least-squares regression takes no account of the continuity of the spectral curves. Jumble up the wavelengths, and continuity is destroyed in Fig. 1.5, and yet out pop exactly the same regression coefficient estimates and predictions. Thus the continuity in Fig. 4.5 must be a consequence of true underlying continuity. A similar phenomenon is present in ridge estimates in Fig. 4.3. Figure 4.7 depicts the cross-validated prediction mean-square error using just 16 of the 1168 frequencies carefully chosen by

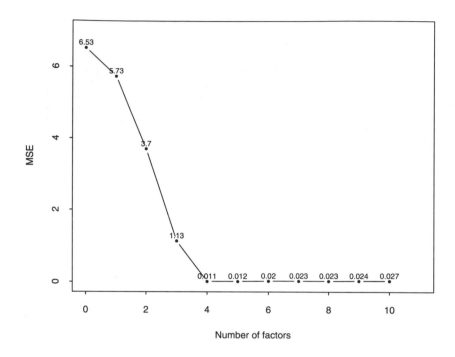

Fig. 4.7. Detergent data, 16 selected wavelengths, cross-validated mean-squared error by number of factors fitted by partial least-squares

the chemist. The same broad shape of curve emerges as in Fig. 4.6, with a minimum of 0.011 at 4 factors this time. However, this minimum is five times smaller than the minimum achieved by including all the 1168 frequencies. We will develop advantageous selection strategies in Chapter 7. These mean-squared errors compare favourably with the ridge regression mean-squared errors for these same detergent data of 0.064. A general Bayesian method is devised in Chapter 6 which takes explicit account of the inherently continuous nature of the spectra as depicted in Fig. 1.5.

4.6.2 PERSPECTIVE ON PARTIAL LEAST-SQUARES

Partial least-squares owes its name to the sequence of simple least-squares calculations that defines it. The model matrix X of n observations on p explanatory variables can be described in bilinear factor form:

$$X = t^{(1)}v_1^T + t^{(2)}v_2^T + \ldots + t^{(\omega)}v_\omega^T + E_\omega$$

where the scores $t^{(j)}$ are n-vectors. They are the latent variables and the p-vectors v_j are the loadings. The residual matrix E_ω is small in some sense. The crucial idea of PLSR is that the relationship between X and y is conveyed through the latent variables. Thus you also have the decomposition

$$y = t^{(1)}b_1 + t^{(2)}b_2 + ...t^{(\omega)}b_\omega + f_\omega$$

for scalar b_j and the *same* scores or latent factors. Incidentally PLSR can quite naturally include multiple y with vector b_j leading to multivariate regression. Various conditions need to be imposed for uniqueness. You could force the scores to be mutually orthogonal in $\mathcal{R}^n$ or the loadings to be mutually orthogonal in $\mathcal{R}^p$. You should avoid imposing both simultaneously since then the $t^{(j)}$ would be eigenvectors of XX^T and v_j eigenvectors of X^TX and the latent factors would be entirely determined by the X data without reference to y, as is the case in the traditional PCR method, which bases choice on eigenvalues. There are thus two main algorithms for PLSR depending on whether the scores or the loadings are determined as orthogonal. The algorithms are sequential, starting with no factors and adding a factor at each stage. For the original algorithm, the orthogonal scores algorithm, latent variables are formed as weighted averages of the X- residuals from the previous step, with weights proportional to covariances with the y-residuals from the previous step (or even just y as in our development from Stone and Brooks (1990)). Only the simplest least-squares algorithms are required. The orthogonal loadings algorithm is superficially rather different and requires a sequence of multiple regressions rather than simple regressions. It turns out to provide a better basis for the development of theoretical properties (see Helland (1988)). Both algorithms lead to the same predictions of y. What emerges in PLSR is that the regression coefficients are formed as in eqn (4.3) where $\mathcal{G}$ is a rank ω approximation to the inverse of X^TX and the usual orthogonal scores algorithm is indeed just the conjugate gradient method of forming an inverse (see Westlake (1968, p. 47)). In PLSR the conjugate directions are formed with respect to y, whereas the traditional form of PCR has approximating inverses on the basis of X^TX alone. PLSR like PCR can be viewed as a shrinkage method, although the shrinkage of PLSR is decidedly more obscure (see Section 4.7) than that of, say, ridge regression, with RR's implicit Bayesian assumption of exchangeability of regression coefficients.

PLSR, like PCR and its close relative, Factor analysis, has this motivational notion of latent factors. This to some extent reduces unease at the 'black box' approach of partial least-squares. We have introduced it as an estimator for model (3.7) but there is really no explicit link to that model. In the statistical linear model there is no principle that directs you to choose latent variables on the basis of maximizing covariance. With this lack of a conventional statistical model goes its consequent lack of attention to prior knowledge encompassed in any substantive application. It offers a biased non-linear shrinkage method but alternatives like ridge regression may be given an explicit Bayesian prior assumption, guiding users as to when and how to use it. It is prescriptive and is not embedded in an inferential framework to judge the relative merits of the prescriptions. Modelling as such is eschewed. In this *soft* science, the chemometrician or statistician user may be left with the comfortable notion that she can collect a batch of data, she does not have to worry too much about how she collects it, or what past knowledge there is. She can apply partial least-squares regression with the assurance that after a bit of fine tuning she will have a good predictor for all future unspecified purposes. The opposing view is that *modelling* is paramount. Whether to regress x on y or y on x would depend on the way the training data has been collected, whether x or y had been controlled. But when $n - 1 < p$ then regressing either way leads to the same least-squares estimates with a $(p - n + 1)$ dimensional degree of indeterminacy. Bayesians would argue that prior information is then crucial in forming a unique estimator.

These caveats should not deflect you from considered use of this technique. It has gained considerable respect and is often effective. The approach to PLSR we have given derives from Stone and Brooks (1990) and is closely related to that of Helland (1988). The Helland algorithm, implemented by Denham (1992) in Fortran embedded in S, and used for computations in this book, is presented as an S listing in the Appendix E: although this is much slower than Fortran embedded in S. Denham also gives comparative timings of algorithms for the implementation of PLSR. For the orthogonal scores and orthogonal loadings algorithms see Martens and Naes (1989). A GENSTAT algorithm for PLSR is given by Rogers (1987).

4.7 Scaling, invariance, and shrinkage

Ordinary least-squares is not changed by rescaling and linearly transforming the x-variables. This is not true of PLSR, PCR, or RR, unless in the case of PLSR and PCR, the maximum possible number of factors are included, when PLSR and PCR are OLS. When $D = X^T X$ is singular, minimum length least-squares as well as the other three methods are affected by scaling. In the case of PCR the maximum number of factors is the rank of D, whereas in PLSR it is at most rank(D) latent factors depending on the number of linearly independent members of the Krylov sequence. However, orthogonal transformations of both the n-rows of observations and the p-columns of variables are allowable for all three techniques without affecting fitted predictions. This is easy to see. Take the model (3.2). Let $X_* = Q^T X P$, $Y_* = Q^T Y$, $\beta_* = P^T \beta$ where Q, $n \times n$ and P, $p \times p$ are orthogonal, then the methods are *invariant* to these transformations if their fitted predictions are unaltered, that is $\hat{Y}_* = Q^T \hat{Y}$, or their parameter estimates are equivalent, that is $\hat{\beta}_* = P^T \hat{\beta}$. Such invariance has already been exploited in deriving properties of ridge regression. In the case of PCR and PLSR note that $D = X^T X$ and $D_* = P^T DP$ and it is immediately evident that PCR is invariant since the eigenvalues of D and D_* are the same, and indeed the orthogonal P consisting of eigenvector of D actually defines the principal components, and is the orthogonal V of the canonical reduction of the regression model leading to eqns (4.8) and (4.9). For PLSR the invariance is not quite so obvious but follows straightforwardly from the covariance maximizations characterization subject to constraints. For example, to generate the first latent factor, maximize $c_*^T d_*$ subject to $c_*^T c_* = 1$, is the same as, maximize $c_*^T P^T d$ subject to $c_*^T c_* = 1$ for which the solution is $c_{1*} = P^T c_1$. Similarly $c_{2*} = P^T c_2$ and so on, and the required invariance follows. For a more formal and less heuristic proof see Denham (1992).

This invariance to orthogonal transformations implies that you may reduce the regression problem to the canonical form, eqns (4.8) and (4.9) of Section 4.3, before applying RR, PCR, or PLSR. This use of the singular value decomposition can help numerical stability and is an aid in understanding and derivation of properties.

If matrices P or Q are not orthogonal then the methods will not generally be invariant. This implies that careful scaling or preprocessing the data can affect the methods described. In the detergent example the comeasureable x-values are not scaled, and scaling was found not advantageous.

A related aspect is the shrinkage properties of the estimators. For ridge regression the squared length of the coefficient p-vector,

$$||\hat{\beta}_{RR}||^2 = ||\hat{\alpha}_{RR}||^2 = \sum \{\lambda_i \hat{\alpha}_i / (\lambda_i + k)\}^2$$

is a monotonic (decreasing) function of k. The minimum length least-squares estimator is achieved when k tends to zero and has squared length as above with $k = 0$. Zero squared length is obtained at $k = \infty$. From the canonical form of the model given by eqns (4.8) and (4.9), a principal components regression estimator has least-squares estimates $\hat{\alpha}_i$ of α_i for $i \in S_\omega$, zero otherwise. Hence the squared length is

$$\sum_{i \in S_\omega} \hat{\alpha}_i^2$$

and is monotone (increasing) in ω, increasing to the MLLS squared length when S_ω, the set of chosen factors, includes all the labels to the non-zero eigenvalues. Since shrinkage occurs on an individual canonical coefficient basis for both ridge and principal components regression it follows that $\sum w_i \hat{\alpha}_{iRR}^2$ shrinks for both these shrinkage estimators. In particular taking $w_i = \lambda_i$ the D-norm or prediction squared length shrinks as k increases

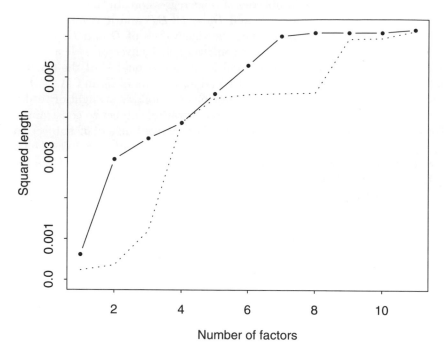

Fig. 4.8. Detergent data, 1168 wavelengths, squared length of regression coefficient estimated vector by number of factors fitted for partial least-squares (solid) and principal components (dotted)

or ω decreases for RR and PCR, respectively. That is, for example, the prediction length is less than or equal to that of least-squares,

$$\hat{y}_R^T \hat{y}_R \le \hat{y}^T \hat{y},$$

where R denotes one of these two regularizing estimators. Since the characterization of partial least-squares is of regression on latent factors, by standard least-squares theory the explained sum of squares is monotone increasing in the number of latent variables fitted and hence for PLSR $\hat{y}_R^T \hat{y}_R = \hat{\beta}_R^T D \hat{\beta}_R$ is increasing in the number of latent variables fitted. It is doubtful but unclear whether Euclidean norm shrinkage applies to partial least-squares, although empirically this seems to hold; see for example the graph of squared length of coefficient vectors for both PLSR and PCR shown in Fig. 4.8.

4.8   Continuum regression

Ordinary least-squares, principal components, and partial least-squares have been tied together in an attractive manner by Stone and Brooks (1990), choosing latent variable(s) by maximizing

$$(c^T d)^2 (c^T D c)^{\gamma - 1}.$$

Here $\gamma = 0$ gives OLS or MLLS regression in one step by maximizing correlation; $\gamma = \infty$ allows variance maximization and corresponds to PCR. In between these two $\gamma = 1$ by maximizing covariance gives the PLSR algorithm described above. Stone and Brooks call this *continuum regression* by virtue of the continuum of values of γ available to define further intervening cases. They use cross-validation to estimate both γ and the number of latent variables. It is not yet clear whether the enriched set of possibilities will in practice lead to better predictions. The methodology suffers from both a heavy computational burden and some inferential opaqueness. What is undoubtedly clear is that, at the very least, it has enhanced understanding of PLSR and the presentation above of PLSR owes much to the Stone and Brooks paper in conjunction with that of Helland (1988) and Sundberg and Brown (1989).

Sundberg (1993) relates CR to RR and argues that CR with $\omega = 1$ should be preferred to RR. He also shows that CR(1) gives estimated regression coefficients that are proportional to the ridge coefficients, where the constant of proportionality is a function of the ridge constant. The proportionality factor is greater than 1, so that the CR(1) coefficients are expanded relative to the RR coefficients. He cites two grounds for preferring CR(1) to RR:

1. CR(1) does not shrink when the design is orthogonal or $p = 1$.

2. CR(1) yields 'as good' predictions as RR (on basis of cross-validatory choice) but with less sensitivity to the choice of γ than to the choice of the ridge constant in RR.

CR(1) for $\gamma < 1$ is easily computed via RR: regress y on the vector of RR fitted values. Rolf Sundberg has since also informed me that this can be extended to $\gamma > 1$ by using negative ridge constants (from $-\infty$ to the largest eigenvalue of D). For CR(2) just repeat the procedure on the residuals from CR(1), and so on.

The issue of good prediction is crucial and we will need to see how this simple reduced form of the method behaves on a variety of datasets. For a Bayesian statistician, the implicit prior structure that would imply the use of CR(1) remains obscure, and its behaviour unsure relative to true underlying parameters.

4.9   Wider considerations

Good data design is of paramount importance in any successful application of regularization methods of this chapter. The design should ideally give a balance by means of an orthogonal design of the main ingredients. Any nuisance factors should be measured and allowed for. Whether they are carefully controlled or allowed to vary depends realistically on how well they can be controlled in real prediction situations away from the careful controls of the laboratory. Are the known impurities or pollutants going to be present in future? Is the temperature to be controlled in future as carefully as it was in the calibration experiment? Probably it is better to contemplate some 'outer' design of the possible nuisance factors, or to make sure that they are varying in a natural manner relative to their real variation.

You may require more than a good estimator in a well controlled situation. You may need some degree of robustness to a wider range of situations and the ability to detect and highlight possibly maverick observations or departures from linearity in the model. Diagnostics for non-linearity are presented in Chapter 7. Prediction diagnostics for outlying future observations are presented in Chapter 5. What PLSR still lacks is a good inferential base. You would like to specify confidence intervals on β, perhaps, and prediction intervals, more surely. PCR and RR are easier to deal with inferentially.

In PLSR it is natural to be a little wary about treating the regression as if the latent factors t are real observations and in the second stage consequently applying standard tests of significance to the b coefficients. It would seem to be better to choose ω by cross-validation and estimate the prediction error from the jackknifed residuals or use the bootstrap method. The bootstrap method resamples from the n observations rather than systematically omitting one observation. In the calibration context, jackknife or bootstrap methods have been developed by Denham (1990). For n as few as 12 though, as in the detergent example, bootstrap methods of providing prediction intervals by sampling the empirical distribution of regression 'errors' have a hard job and you can do little better than the naive approach of using variance (4.17) with $\mathcal{G}$ defined by PLSR. This is tantamount to regarding the latent variables in eqn (4.22) as given *a priori*. Furthermore, accumulating empirical evidence in a variety of examples suggests that despite reservations, regarding latent variables as if fixed is often adequate.

The example as described is univariate, involving one composition variable, Y. With several compositional variables you can either treat them separately or apply multivariate generalizations of the univariate techniques. For multivariate versions of RR see Brown and Zidek (1980). Multivariate PLSR demands common latent variables for the different compositions and this may not always be desirable.

Algorithms to implement PLSR are very simple and can with advantage be incorporated into many statistical systems. Algorithms for PCR and RR are even simpler and can be based on the singular value decomposition. Computations for cross-validatory choice in continuum regression have the heaviest computational burden. We have used S (see Becker *et al.* (1988)), on a SUN workstation to do the analysis and graphics of this chapter and the rest of the book. A more portable but similar object-oriented approach is provided by LISP-STAT (Tierney (1990)). Whilst not incorporating all the features of S, it is especially good at dynamic graphics. This enables you to explore higher dimensional point clouds and surfaces. Other systems such as SAS, GENSTAT, or GAUSS on a PC could also be used as a wider system for data handling, graphing, and general exploratory data analysis (see Payne *et al.* (1988) for a critique of the methods for incorporating new statistical software). Scientists may prefer a more tailored package but should be aware that this may entail some loss of flexibility and control and some possible sacrifice of understanding when it comes to interpreting the output.

5
Multivariate calibration

5.1 Introduction

Multivariate regression is concerned with the relationship between two sets of variables. The two sets of variables are treated asymmetrically. The response set is regarded as random and the explanatory set treated as if it is fixed, even though in reality both may be random. Inference may centre on predicting the random set given values of the fixed set, and this is standard regression prediction; or it may concentrate on estimation of the fixed explanatory vector that gave rise to observations on the random set, the calibration problem. Multiple regression in the previous two chapters dealt with the special standard case of just one response variable. Just as in this univariate case, the multivariate calibration problem depends crucially on whether in reality in the training data the explanatory variables are randomly observed or are fixed by design.

   Here the emphasis is on standard normal theory but techniques may in principle be extended to other parametric forms of distribution. Classical, likelihood, and Bayesian solutions are presented. For a discussion of the range of more standard statistical tests for model validity the reader is referred to treatises such as Mardia *et al.* (1979), or the exploratory data analysis of Gnanadesikan (1977). Somewhat non-standard multivariate distributional notation following Dawid (1981) is adopted throughout this and subsequent chapters; for details see Appendix A. This notation has considerable advantages over the more standard notation both in ease of manipulation, conforming degrees of freedom under marginalization, and extension to doubly countable matrices.

5.2 Classical prediction

In this section the prediction of a future $Z(q \times 1)$ at a given value of the explanatory vector (x) is considered. This provides sampling theory distributional results for use in later sections when x is unknown (and then designated ξ). If m replicates are available at the same x then the predictive distribution of $\bar{Z}$ is developed.

   The assumed multivariate linear regression model for the n-sample training data is

$$Y = 1_n \mu^T + XB + \mathcal{E} \tag{5.1}$$

where $Y(n \times q), \mathcal{E}(n \times q)$ are random matrices, $X(n \times p)$ is a fixed matrix of explanatory variables each centred as eqn (3.6), 1_n is a vector of ones and $\mu(q \times 1)$ and $B(p \times q)$ are unknown parameters. The error matrix $\mathcal{E}(n \times q)$ is such that with $\mathcal{E} = (\epsilon_1, \ldots, \epsilon_n)^T$,

$$E(\epsilon_i) = 0, \quad E(\epsilon_i \epsilon_i^T) = \Gamma, \quad E(\epsilon_i \epsilon_j^T) = 0, \tag{5.2}$$

for $i, j = 1, \ldots, n$. These are additionally multivariate normal. The model for m independent future further response vectors each observed independently at the same set of explanatory variables, x, is

$$Z = 1_m \mu^T + 1_m x^T B + \mathcal{E}^* \tag{5.3}$$

where the m rows of the error matrix $\mathcal{E}^*(m \times q)$ satisfy (5.2) and are normal and independent of $\mathcal{E}$. Since eqn (5.3) consists of m independent observations with the same mean, you may reduce to sufficient statistics $\bar{Z}$ and $m - 1$ independent contrasts with mean vector zero and covariance matrix Γ. We are typically interested in predicting $\bar{Z}$. The distribution of $\bar{Z}$ is independent of the distribution of the least-squares $\hat{\mu}$ and $\hat{B}$ from the training data and model (5.1). Least-squares estimation of the parameters, μ and B, of model (5.1) is identical to q separate univariate multiple regression least-squares estimators,

$$\hat{\mu} = \bar{Y}, \quad \hat{B} = (X^T X)^{-1} X^T Y \tag{5.4}$$

and under normality $\hat{\mu}$ and $\hat{B}$ are independent,

$$\hat{\mu}\sqrt{n} - \mu\sqrt{n} \sim N_q(0, \Gamma);$$

and in the matrix normal notation of Appendix A,

$$\hat{B} - B \sim \mathcal{N}(G, \Gamma),$$

where the $(p \times p)$ matrix
$$G = (X^T X)^{-1} \tag{5.5}$$

expresses the covariance across rows, $\Gamma(q \times q)$ by columns. It is straightforward to use a canonical form of model (5.1) as in eqns (4.8) and (4.9) to show

$$(\bar{Z} - \hat{\mu} - \hat{B}^T x)/c_{m,n}(x) \sim N_q(0, \Gamma),$$

where the scalar

$$c_{m,n}^2(x) = 1/m + 1/n + x^T (X^T X)^{-1} x. \tag{5.6}$$

Let $S(q \times q)$ be the residual sum of products matrix from the training data,

$$S = (Y - 1_n \hat{\mu}^T - X\hat{B})^T (Y - 1_n \hat{\mu}^T - X\hat{B}). \tag{5.7}$$

This will later be pooled with the prediction contrast data when Z is actually observed. Now S has a Wishart distribution with scale matrix Γ and degrees of freedom $\nu + q - 1$ (see Appendix A) where

$$\nu = n - p - q.$$

From the distributional conventions and characterizations of Appendix A, it follows that for $\nu > 0$,

$$S^{-1/2}(\bar{Z} - \hat{\mu} - \hat{B}^T x)/c_{m,n}(x) \sim \mathcal{T}(\nu; I_q), \tag{5.8}$$

where $S = S^{1/2}(S^{1/2})^T$ and $S^{1/2}$ is a square root matrix of S; and a $100(1 - \gamma)$ per cent prediction ellipsoid for the future $\bar{Z}$ is, from Appendix A.6,

$$(\bar{Z} - \hat{\mu} - \hat{B}^T x)^T S^{-1}(\bar{Z} - \hat{\mu} - \hat{B}^T x)/c_{m,n}^2(x) \leq (q/\nu)F_\nu^q(\gamma), \tag{5.9}$$

with $F_\nu^q(\gamma)$ the upper $100(1 - \gamma)$ percentage point of the standard F-distribution on q and ν degrees of freedom. This reduces to the form of earlier chapters when $q = 1$ with either $p > 1$ or $p = 1$. The sampling results refer to repetitions of both $Y(n \times q)$ and $\bar{Z}$ conditional on fixed, given $X(n \times p)$ and $x(p \times 1)$, although since the resultant distributions do not involve X or x, the statements are also true unconditionally.

5.3   Controlled calibration

The previous section was concerned with prediction of $\bar{Z}$ for observed $x(p \times 1)$. From observed $Z(m \times q)$ you may wish to estimate the unknown explanatory vector, designating this unknown x as ξ. Special inferential problems now arise that were not present in earlier chapters. Suppose $q \geq p$ as is often the case. Difficulties arise when strictly $q > p$. Then the dimension of the observed $\bar{Z}$ is greater than the dimension of the unknown ξ which makes up its mean. The model (5.1), error structure (5.2) with normal errors, together with the unknown ξ version of prediction model (5.3) gives

$$Z = 1_m \mu^T + 1_m \xi^T B + \mathcal{E}^*, \tag{5.10}$$

with the same error structure. It forms an exponential family which is curved when $q > p$ (see Section 5.12). Likelihood and Bayesian inference behave quite reliably whether or not the exponential family is curved.

Classical sampling inference can be quite misleading without careful considerations of ancillarity. In this section the difficulties that may arise are explored. Bayesian and likelihood inference are covered in later sections.

Firstly from eqn (5.10) you may reduce to sufficient statistics, $\bar{Z}$ and S' where

$$S' = (Z - 1_m \bar{Z}^T)^T (Z - 1_m \bar{Z}^T) \tag{5.11}$$

is the error sum of products matrix on $m - 1$ degrees of freedom. Provided you are satisfied that the error from eqn (5.10) has the same covariance structure as eqn (5.2) of the training model, then you can add the two error sum of products matrices, S and S', to form

$$S_+ = S + S' \tag{5.12}$$

with combined degrees of freedom, $\nu + q - 1$, where now

$$\nu = n - p - q + m - 1. \tag{5.13}$$

From the development leading to eqn (5.9) a $100(1 - \gamma)$ per cent confidence region for ξ is all ξ such that

$$(\bar{Z} - \hat{\mu} - \hat{B}^T \xi)^T S_+^{-1} (\bar{Z} - \hat{\mu} - \hat{B}^T \xi)/c_{m,n}^2(\xi) \leq (q/\nu) F_\nu^q(\gamma). \tag{5.14}$$

The generalized least-squares estimator may be derived from the model for $\bar{Z}$ of model (5.10), treating all parameters as if known to be their least-squares plugged-in values. It also minimizes the numerator of the left-hand side of eqn (5.14), and is

$$\hat{\xi} = K^{-1} \hat{B} \hat{\Gamma}^{-1} (\bar{Z} - \hat{\mu}) \tag{5.15}$$

where

$$K = \hat{B} \hat{\Gamma}^{-1} \hat{B}^T. \tag{5.16}$$

Here the estimator of Γ is taken to be

$$\hat{\Gamma} = S_+/(\nu + q - 1), \tag{5.17}$$

although the purpose is not to estimate Γ and any divisor of S_+ leaves the the estimator $\hat{\xi}$ unchanged. This means that the true covariance matrix from model (5.1) may differ from that of the prediction model (5.10) by a pure scale change, and yet leave estimator $\hat{\xi}$ unchanged, although confidence regions would be affected. The confidence region (5.14) may be written

$$(\|\xi - \hat{\xi}\|_K^2 + R)/c_{m,n}^2(\xi) \leq q F_\nu^q(\gamma), \tag{5.18}$$

where
$$R = ||\bar{Z} - \hat{\mu} - \hat{B}^T \hat{\xi}||^2_{\hat{\Gamma}^{-1}} \qquad (5.19)$$

is the residual variation in $\bar{Z}$ after explanation by $\hat{\xi}$ in the norm $\hat{\Gamma}^{-1}$. Also R may be interpreted as a diagnostic measuring discrepancies between the q components of $\bar{Z}$ as to their implied estimate of ξ. It is identically zero if the number of responses and explanatory variables are equal. Otherwise since
$$\Gamma^{-1/2}(\bar{Z} - \mu - B^T \xi) \sqrt{m} \sim \mathcal{N}(I_q, 1),$$

mR asymptotically, as $n \to \infty$, has a chi-squared distribution on $(q - p)$ degrees of freedom. For $m = 1$, asymptotic expansions for R, taking account of uncertainty in estimates of μ, B, and Γ, are given by Davis and Hayakawa (1987). To a first order

$$R/c^2_{1,n} \sim (q - p)F^{q-p}_\nu.$$

The region of ξ values that satisfies the inequality (5.18) is monotonically decreasing in R. As R increases the region will shrink to a point and then become empty. This unsatisfactory behaviour sullies the procedure's exact confidence property. The more contradictory the observed $\bar{Z}$ the more restricted the confidence region. Admittedly such large values of R have low probability under model (5.1), (5.10), but at the very least the discrepant behaviour should be highlighted rather than the model relied on to pin down ξ more precisely. Indeed Williams (1959) suggests the use of R as a test of agreement between the components of $\bar{Z}$.

Let us look at inequality (5.18) more closely. It may be rationalized and written as a quadratic form in ξ. Let

$$k = qF^q_\nu(\gamma).$$

When $q = p$ and consequently $R = 0$, the squared term is $\xi^T C \xi$, with

$$C = K - kG,$$

and G and K given by eqns (5.5) and (5.16) respectively. Then C positive definite ensures that the region is a bounded ellipsoid and an interval when $q = p = 1$, conforming with the predictive Student procedure of Chapter 2. Otherwise when $R > 0$ not even this condition on C is sufficient to guarantee a bounded region. Further details are given in Brown (1982). The difficulties with this exact confidence procedure have also been noticed in practice by Wood (1982) and Oman and Wax (1984). Wood's preferred solution is to ignore R and base confidence procedures on the distribution of the generalized least-squares estimator (5.15); see also Fujikoshi and

Nishii (1984) and Davis and Hayakawa (1987) for detailed distributional results. Oman (1988) develops a translation invariant procedure which also projects so as to ignore R. Both likelihood and Bayes procedures discussed in this chapter essentially behave as if R is an approximate ancillary with the ancillary determining the precision of estimation in such a way that confidence regions *widen* with increasing R.

5.4   Random calibration

Again suppose inference concerning an unobserved explanatory vector ξ is of interest. However, in the training data the explanatory x-variables have not been controlled but vary randomly. What's more the random generating mechanism is supposed similarly to hold for the as yet unobserved vector ξ. This is what has been termed random or natural calibration in earlier chapters. For the solution to the inference problem you may simply reverse the roles of X and Y and use the conditional distribution of X given Y to predict ξ. In terms of least-squares regression estimation, the 'inverse' regression predictor, $\check{\xi}$, is given as

$$\check{\xi} - \bar{X} = S_{xy}S_{yy}^{-1}(Z - \bar{Y}), \qquad (5.20)$$

where S_{xy} and S_{yy} are sums of products matrices from the centred n-observation training data,

$$S_{xy} = X^TY, \ S_{xx} = X^TX, \ S_{yy} = Y^TY,$$

when $X(n \times p)$ and $Y(n \times q)$ have been centred so that column totals are zero. Notice that predictor $\check{\xi}$ with m replicates and $\bar{Z}$ replacing Z is not sensible, as remarks around formula (2.28) discuss. For this case, a modified estimator may be derived in the same way as for simple linear regression.

One can also provide a prediction ellipsoid for vector ξ, essentially by applying Section 5.2, and reversing the roles of X and Y. Here there is no problem with the behaviour of the regions. This is because eqn (5.20) is based on the stronger assumptions of natural calibration, utilizing a p-variate multivariate normal distribution for the n x-vectors of the training data and ξ. This will also be evident from the differing implicit Bayesian roots of the two estimators. In the next section the two estimators are compared.

5.5   Comparison of two regressions

Firstly comparative formulae for regressing Y on X and X on Y are developed; that is, relating the least-squares estimators $\hat{\xi}$ and $\check{\xi}$, respectively. For $p = q = 1$ the two are proportional with constant of proportionality the square of the correlation, as shown in Section 2.3. For general $p \leq q$ it transpires that after a certain $(p \times p)$ matrix transformation applied to each, the two estimators are again proportional (elementwise) with constants of proportionality the squared canonical correlations. This relationship holds also for singular cases except that for $n = p + q + 1 - j, j = 1, \ldots, p$ there are j canonical correlations of unity, and for $p \leq n \leq q$ all canonical correlations are unity and the two estimators are *identical* and free to vary in the *same* indeterminate subspace of dimension $(q - n + 1)$. The formulae for the non-singular case are developed below. The singular case identities follow most easily from a canonical reduction detailed in Sundberg and Brown (1989). Secondly in this section moment results for the two estimators are surveyed.

The development of comparative formulae in the non-singular case is a simplified version of Brown (1982), suggested to me by Timo Mäkeläinen. The columns of both X and Y have been centred. From estimator (5.20) let $\check{\xi} = AZ$ where the $(p \times q)$ matrix A is given as

$$
\begin{aligned}
A &= X^T Y (Y^T Y)^{-1} \\
&= X^T X \hat{B}(S + \hat{B}^T X^T X \hat{B})^{-1},
\end{aligned}
$$

where the above uses eqn (5.7). Further manipulation gives

$$
\begin{aligned}
A &= X^T X \hat{B}(I + S^{-1}\hat{B}^T X^T X \hat{B})^{-1}S^{-1} \\
&= X^T X (I + \hat{B}S^{-1}\hat{B}^T X^T X)^{-1}\hat{B}S^{-1}, \tag{5.21}
\end{aligned}
$$

where the last step follows from

$$
F(I + EF)^{-1} = (I + FE)^{-1}F,
$$

itself a consequence of $(I + FE)F = F(I + EF)$. Combining this with eqn (5.15), in the version with S in place of S_+, the linking relationship is

$$
\check{\xi} = X^T X (I + \hat{B}S^{-1}\hat{B}^T X^T X)^{-1}(\hat{B}S^{-1}\hat{B}^T)\hat{\xi}, \tag{5.22}
$$

a matrix weighted average between $\hat{\xi}$ and the zero p-vector. Using again eqn (5.21), but backwards this time,

$$
\begin{aligned}
\check{\xi} &= X^T Y (Y^T Y)^{-1}\hat{B}^T \hat{\xi} \\
&= (X^T X)^{1/2}UU^T (X^T X)^{-1/2}\hat{\xi},
\end{aligned}
$$

or

$$
(X^T X)^{-1/2}\check{\xi} = UU^T (X^T X)^{-1/2}\hat{\xi}, \tag{5.23}
$$

where

$$U = (X^TX)^{-1/2}X^TY(Y^TY)^{-1/2},$$

so that the positive singular values of U are the usual canonical correlations between X and Y.

In summary, after the same particular non-singular transformation, $Q^T(X^TX)^{-1/2}$, where columns of Q are orthonormal latent vectors of UU^T, components of the transformed $\check{\xi}$ and $\hat{\xi}$ are simply proportional, the p constants of proportionality being the p squared canonical correlations between X and Y. This confirms the sense in which $\check{\xi}$ is a shrunken version of $\hat{\xi}$, and generalizes the comparison for single X and Y of Section 2.3.

The relationship (5.22) may be used as a basis for comparison of the two estimators as regards mean, variance, and mean-squared error under the controlled calibration model. Whereas estimator (5.15) is asymptotically unbiased and estimator (5.20) biased, the shrunken (5.20) has 'smaller' variance, so that the asymptotic mean-squared error of estimator (5.20) is smaller than that of estimator (5.15) over a central region of ξ values, although generally the improvement is small and does not outweigh more general considerations which would favour the use of the classical estimator in controlled calibration. Some details are given below. Small sample comparisons are more problematic. Firstly, whereas all moments of the inverse estimator (5.20) exist this is not true of the classical estimator. Nishii and Krishnaiah (1988) show that provided $n \geq \max\{q+3, p+q+1\}$ then

1. the mean vector of $\hat{\xi}$ is finite if and only if $q \geq p+1$,
2. the mean-squared error of $\hat{\xi}$ is finite if and only if $q \geq p+2$.

The same result with $p = 1$ is given by Brown and Spiegelman (1991) in the wider context of selection of responses. These necessary and sufficient conditions sharpen the sufficient conditions of Lieftinck-Koeijers (1988). For $p = 1$ both she and Nishii and Krishnaiah also give formulae for the mean and mean-squared error in terms of moments of an inverse linear function of Poisson random variables.

The moments of the asymptotic distribution are finite and a large sample or, rather, small error comparison of the estimators under model (5.1), (5.10) offers the simplest comparison. For this comparison, B is substituted for $\hat{B}$ and Γ for $\hat{\Gamma}$, thereby neglecting contributions from uncertainty in estimators of B and Γ. This may be justified by noting that in most practical applications practitioners aim to make the calibration precise enough for error in the calibration function to be of minor importance. Quoting Berkson (1969): 'A calibration line is, almost by definition, without appreciable error'. Thus asymptotically the prediction model may be considered as if it has known parameters, from which

1. $E(\hat{\xi}) = \xi$,
2. $Cov(\hat{\xi}) = (B\Gamma^{-1}B^T)^{-1}$;
3. $E(\check{\xi}) = (I_p - \mathcal{R})\xi$,
4. $Cov(\check{\xi}) = (I_p - \mathcal{R})Cov(\hat{\xi})(I_p - \mathcal{R})^T$,

where from eqn (5.22)

$$\mathcal{R} = (I_p + \Sigma B\Gamma^{-1}B^T)^{-1}$$

and

$$\Sigma = X^T X/n, \tag{5.24}$$

is the covariance matrix of the x-variables, X being centred throughout.

The mean-squared error comparison is probably best effected through the canonical transformation given above. This allows direct comparison with the simple linear regression results of Section 2.6. For a similar analysis without the canonical transformation see Sundberg (1985). Let $\eta = Q^T\Sigma^{-1/2}\xi$, $\hat{\eta} = Q^T\Sigma^{-1/2}\hat{\xi}$, $\check{\eta} = Q^T\Sigma^{-1/2}\check{\xi}$, where columns of Q are eigenvectors $\Sigma^{1/2}B\Gamma^{-1}B^T\Sigma^{1/2}$. Note that if ξ is 'like' x of calibration in the sense that $\xi \sim \mathcal{N}(\Sigma, 1)$, then after the transformation,

$$\eta \sim \mathcal{N}(I, 1). \tag{5.25}$$

The relationship becomes componentwise

$$\check{\eta}_j = \rho_j^2\hat{\eta}_j,$$

and

$$E(\hat{\eta}_j) = \eta_j, \quad \mathrm{Var}(\hat{\eta}_j) = (1 - \rho_j^2)/\rho_j^2;$$

$$E(\check{\eta}_j) = \rho_j^2\eta_j, \quad \mathrm{Var}(\check{\eta}_j) = \rho_j^2(1 - \rho_j^2) \; j = 1, \ldots, p.$$

Hence the mean-squared error of $\hat{\eta}$ subtracted from that of $\check{\eta}$ is

$$\{\rho_j^2(1 - \rho_j^2) + \eta_j^2(1 - \rho_j^2)^2\} - (1 - \rho_j^2)/\rho_j^2$$

so that $\check{\eta}$ offers an improvement if and only if

$$\eta_j^2 \leq 1 + 1/\rho_j^2. \tag{5.26}$$

This is exactly the same componentwise as the univariate condition (2.31). The right-hand side of this inequality is at least 2. Also, by the Berkson maxim ρ_j^2 will be close to 1 for $j = 1, \ldots, p$, so that this bound of 2 may be not so far from the typical value. For a chosen component of η, under distributional assumption (5.25) the probability of satisfying $\eta_j^2 \leq 2$ is that of a chi-squared random variable on one degree of freedom being less than

2. This is 0.84. The probability of this holding for all p components is $(0.84)^p$ and will be much smaller for larger numbers of components.

In reality the calibration will not provide perfectly accurate estimates of the parameters. Fujikoshi and Nishii (1986) give asymptotic expansions up to terms of order $\frac{1}{n}$ for both the mean and a (weighted) mean-squared error of $\hat{\xi}$. For the mean of $\hat{\xi}$ they give

$$E(\hat{\xi} - \xi) = -\frac{1}{n}(q - p - 1)(B\Gamma^{-1}B^T)^{-1}\Sigma^{-1}\xi + o\left(\frac{1}{n}\right).$$

The order $\frac{1}{n}$ bias above can be substantial when (a) n is small, (b) q is large. The covariance matrix of $\hat{\xi}$ will also be strongly affected in these circumstances, primarily because of poor estimation of Γ in estimator (5.15). See a later Bayesian remedy to this and further discussion in Section 5.12.

5.6   Bayesian prediction

We first describe a Bayesian analysis of model (5.1) for subsequent prediction of $\bar{Z}$ in model (5.10), with X centred, and independent $N_q(0, \Gamma)$ errors. Under this model the unknown parameters are, $\mu, B, \Gamma,$ and ξ. It is probably natural to assume that *a priori* ξ, the unknown explanatory vector, is independent of the model parameters, $\mu, B,$ and Γ :

$$\pi(\mu, B, \Gamma, \xi) = \pi(\mu, B, \Gamma)\pi(\xi). \tag{5.27}$$

First, you may form a posterior distribution on $\mu, B,$ and Γ solely from model (5.1), (5.2). The likelihood function is

$$p(Y|X, \mu, B, \Gamma) \quad \propto \quad |\Gamma|^{-n/2}\exp[(-1/2)\text{trace}\Gamma^{-1}\{S + n(\mu - \hat{\mu})(\mu - \hat{\mu})^T$$
$$+(B - \hat{B})^T X^T X (B - \hat{B})\}] \tag{5.28}$$

where (ˆ) denotes least-squares estimators, which after centring X are given by eqn (5.4); and the residual sum of products, S, which will be used to estimate Γ, is given by eqn (5.7). This likelihood function is of the form

$$f(\Gamma)g(\mu, B|\Gamma),$$

and the particular functional forms suggest a natural conjugate prior distribution of the following form. With the distributional notation of Appendix A, given Γ,

$$\Theta - A \sim \mathcal{N}(C, \Gamma)$$

where $\Theta^T = (\mu, B^T)$; and marginally

$$\Gamma \sim \mathcal{IW}(d; D),$$

and scalar d and matrices $A\,[(p+1)\times q]$, $C\,[(p+1)\times(p+1)]$, and $D\,(q\times q)$ are prespecified by prior knowledge. This is the natural conjugate Normal–Inverse–Wishart distribution. The prior matrices may involve further hyperparameters which have further prior assumptions forming a hierarchy of priors akin to that of Lindley and Smith (1972). This progression of prior layers may stop either with specified hyperparameters or with 'vague' prior distributions. Chen (1979) allows one layer of hyperparameters and these are then specified with uninformative prior distributions. The results of this chapter allow informative prior distributions, although vague Jeffreys invariant priors are ultimately usually assumed for the original parameters μ, B, and Γ, so as to provide a baseline analysis with results of similar form to classical and likelihood forms. Informative natural conjugate priors and other forms arise explicitly in the more advanced approaches in Chapter 6 in the context of a continuum of variables. From the perspective of flexible assignment of prior distributions the natural conjugate Normal–Inverse–Wishart has serious limitations, primarily stemming from the single scalar, d, which specifies uncertainty in the prescribed covariance matrix D. Brown *et al.* (1993a) define the *generalized inverse Wishart* which provides an enrichment of the inverse Wishart to allow different precision on sections of the covariance matrix. Even with informative prior distributions, μ, the mean of the Y, and B, the regression coefficients, are treated quite separately usually, and taken to be *a priori* independent given Γ, with perhaps an informative prior on Γ only. This at least is necessary if the data are relatively uninformative with $n-p-q-1$ less than or close to zero. Results with natural conjugate informative prior distributions may be derived from Appendix B.

For the uninformative prior specification, it is assumed that the data are informative, that is $n - p - q - 1 \geq 0$. The invariant Jeffreys prior for μ, B, and Γ is

$$\pi(\mu, B, \Gamma) \propto |\Gamma|^{-(q+1)/2}. \qquad (5.29)$$

Combining this with eqn (5.28), the posterior distribution is such that, independently given Γ,

$$(\mu - \hat{\mu})^T \sim \mathcal{N}(1/n, \Gamma), \quad (B - \hat{B}) \sim \mathcal{N}(G, \Gamma), \qquad (5.30)$$

with $\hat{\mu}$ and $\hat{B}$ given by eqn (5.4) and G by eqn (5.5). To verify this note that the density functions of the above involve stripping out from eqn (5.28) terms $|\Gamma|^{-1/2}$ for μ and $|\Gamma|^{-p/2}$ for B as a multipliers of the exponential term. In the case of B this follows from writing B in vector form when the pq-vector has covariance matrix $\Gamma \otimes G$ with determinant $|\Gamma|^p|G|^q$. Hence the the power of $|\Gamma|$ in the marginal Inverse Wishart kernel is $-(n+q-p)/2$ which implies that

$$\Gamma \sim \mathcal{IW}(\nu; S), \qquad (5.31)$$

in the Dawid distributional notation of Appendix A. Here the shape parameter is

$$\nu = (n + q - p) - 2q = n - p - q,$$

as in eqn (5.13), but with $m = 1$ since the left-hand side of eqn (5.10) is to be predicted and is not available as data.

In summary, in the normal error model (5.1), (5.2), with invariant Jeffreys prior (5.29) the posterior distribution is a Normal–Inverse–Wishart with, given Γ, independent normals for μ and B given by (5.30) and marginally Γ has an Inverse Wishart distribution with shape parameter ν, and scale matrix S, all distributions defined by Appendix A.

Now it is straightforward to derive the predictive distribution of $\bar{Z}$ from the m repetitions of model (5.10). You could also predict the $m - 1$ contrasts, but since their distribution does not involve (μ, B) that is seldom of interest; once observed S' given by eqn (5.11) contributes $m - 1$ further degrees of freedom to inference concerning Γ.

Given μ, B, and Γ,

$$\bar{Z} - \mu - B^T x \sim \mathcal{N}(\Gamma, 1/m).$$

Taken with eqn (5.30) and the additive property of independent normal random vectors, given Γ,

$$\bar{Z} - \hat{\mu} - \hat{B}^T x \sim \mathcal{N}(\Gamma, c_{m,n}^2(x)),$$

with $c_{m,n}(x)$ given by eqn (5.6). Using eqn (5.31) and the characterization of a multivariate T-distribution, the predictive distribution of

$$(\bar{Z} - \hat{\mu} - \hat{B}^T x)/c_{m,n}(x) \tag{5.32}$$

is multivariate Student $\mathcal{T}(\nu; S)$. This in turn implies that the sampling distributional result (5.9) holds also with a Bayesian predictive interpretation after an uninformative Jeffreys prior. Predictive distributions are used extensively in Bayesian discrimination, as discussed in Chapter 8.

5.7 Bayesian calibration

In this section training data from model (5.1) and prediction data from model (5.10) are incorporated in a Bayesian analysis. The formulation is initially for quite general prior distributions and then later specializes to an invariant Jeffreys prior for the parameters (μ, B, Γ). Informative natural conjugate priors are easy to accommodate. In this section both uninformative and informative prior assumptions for the future explanatory vector, ξ, are assumed within the assumed independence structure (5.27). A vague prior for ξ is commensurate with controlled calibration whereas a very particular informative prior which treats ξ as exchangeable with the x of the training data leads, when $m = 1$, exactly to a random calibration formulation.

The main result is presented as a theorem which forms a straightforward application of Bayes theorem. The assumptions of the theorem are:

- the model is defined by eqns (5.1), (5.2), (5.10), and normality
- the prior distribution satisfies the independence assumption (5.27)
- let Z be split into the normal sufficient statistics, $\bar{Z}$ and S', defined by eqn (5.11)

Theorem 5.1. *With the assumptions above, the posterior distribution of ξ given $X, Y,$ and Z is*

$$\pi(\xi|X,Y,Z) \propto L(\xi|X,Y,Z)\pi(\xi|X), \tag{5.33}$$

where $L(\xi|X,Y,Z)$ is the predictive likelihood of $\bar{Z}$, including S' when $m > 1$, that is

$$L(\xi|X,Y,Z) = p(\bar{Z}|X,Y,S',\xi),$$

where the density $p(.|.)$ from eqn (5.10) has been marginalized over the posterior distribution of parameters (μ, B, Γ) given $X, Y,$ and S'.

Proof. Dropping factors not involving ξ, indicating this with a proportionality sign, and freely using conditional probability: $p(A, B) = p(A|B)p(B) = p(B|A)p(A)$,

$$
\begin{aligned}
\pi(\xi|X,Y,Z) &= \pi(\xi|X,Y,\bar{Z},S') \\
&\propto \pi(Y,\bar{Z}|\xi,X,S')\pi(\xi|X,S') \\
&= \pi(Y,\bar{Z}|\xi,X,S')\pi(\xi|X) \\
&= \pi(\bar{Z}|Y,\xi,X,S')\pi(Y|\xi,X,S')\pi(\xi|X) \\
&\propto \pi(\bar{Z}|Y,\xi,X,S')\pi(\xi|X),
\end{aligned}
$$

which is the required result.

Remark. When $m = 1$, the result applies to non-normal error distributions.

We now specialize to the Jeffreys invariant prior on $\mu, B,$ and Γ given by eqn (5.29). Let S_+ be the augmented sum of products matrix of errors, in eqn (5.12), with combined degrees of freedom, ν, in eqn (5.13). The required predictive distribution of $\bar{Z}$ is such that eqn (5.32) is distributed as $\mathcal{T}(\nu; S_+)$. Utilizing the density of the multivariate T-distribution given by Appendix formula (A.5), as a function of ξ,

$$L(\xi|X,Y,Z) \propto \{c^2_{m,n}(\xi)\}^{\nu/2}/\{c^2_{m,n}(\xi)+(\bar{Z}-\hat{B}^T\xi)^T S_+^{-1}(\bar{Z}-\hat{B}^T\xi)\}^{(\nu+q)/2}, \tag{5.34}$$

where both X and Y are centred in the training data and $\bar{Z}$ is measured on this Y-centred scale. To adopt an uncentred Y-scale replace $\bar{Z}$ by $\bar{Z} - \hat{\mu}$. This predictive or Bayes integrated likelihood is the ratio of two quadratic forms in ξ raised to powers $\nu/2$ and $(\nu + q)/2$. For large $||\xi||$ it behaves like $1/||\xi||^q$ and is integrable provided $q \geq 2$. When $q = 1$ as in Hoadley (1970), a proper posterior distribution $\pi(\xi|X)$ is necessary in eqn (5.33) for overall integrability. Note that following eqn (5.18), eqn (5.34) may be written as

$$\left(\frac{1}{m} + \frac{1}{n} + ||\xi||_G^2 \right)^{-q/2} \left[1 + \frac{1}{\nu}\{R + ||\xi - \hat{\xi}||_K^2\}/(\frac{1}{m} + \frac{1}{n} + ||\xi||_G^2) \right]^{-(\nu+q)/2}$$

$$(5.35)$$

where $G, \hat{\xi}, K$, and R are given by eqns (5.5), (5.15), (5.16), and (5.19), respectively. Firstly with $R = 0$, if it were not for the first factor, eqn (5.35) would be maximized at $\xi = \hat{\xi}$. The first factor tends to shift this maximum towards the origin but the effect will be slight as ν increases relative to q. The effect of $R > 0$ is to shift the maximum away from the origin. The behaviour of eqn (5.35) could be investigated by simultaneously diagonalizing G and K; that is, it depends on the eigenvalues of $G^{-1}K$ or UU^T of Section 5.5; in other words the canonical correlations between X and Y.

If X is informative about ξ then $\pi(\xi|X)$ will have a crucial role in modifying the predictive likelihood to form the overall posterior distribution. One special form of $\pi(\xi|X)$ is particularly instructive in showing the influence of prior knowledge and the structure of the relationship between X and Y.

For the following theorem from Brown (1982) assume that there are no replicates ($m = 1$). It shows that if ξ is 'like' rows of X, then the posterior distribution of ξ is as if the roles of X and Y had been reversed and X regressed on Y. Specifically, suppose *a priori* that ξ and rows of uncentred $X(n \times p)$ form independent samples from a $N_p(\theta_2, \Sigma_{22})$, that is ξ is exchangeable with explanatory row vectors of the training data. The vague prior

$$\pi(\theta_2, \Sigma_{22}) \propto |\Sigma_{22}|^{-(1-q)/2}, \qquad (5.36)$$

gives the posterior distribution of ξ given X as Student $\mathcal{T}(\nu - p; (1 + 1/n)X^TX)$.

Theorem 5.2. *Suppose given X that $\xi \sim \mathcal{T}(\nu-p; (1+1/n)X^TX)$, then the posterior distribution of ξ is $\check{\xi} + \mathcal{T}(\nu+q-p; \{1+1/n+Z^T(Y^TY)^{-1}Z\}(G+K)^{-1})$, with X and Y centred.*

Proof. From the prior assumption of Theorem 5.2

$$\pi(\xi|X) \propto (1 + 1/n + \xi^T(X^TX)^{-1}\xi)^{-\nu/2} = (c_{1,n}(\xi))^{-\nu}$$

and this term knocks out the numerator in the version of eqn (5.34) with $m = 1$, leaving the denominator as the power of a quadratic form in ξ, the kernel of a multivariate T density. Completing the square in this quadratic form, after a little algebra the result follows.

Remark 1. The assumed prior of eqn (5.36) for θ_2 and Σ_{22}, although not the Jeffreys invariant prior, is correct for creating the joint posterior predictive distribution of (ξ^T, Z^T) given X and Y when the joint distribution of the $(p+q)$ variables arises from sampling from $N_{p+q}(\theta, \Sigma)$ with prior

$$\pi(\theta, \Sigma) \propto |\Sigma|^{-(q+1)/2}.$$

This may be seen from evaluating the Jacobian of the transformation from

$$(\theta, \Sigma) \rightarrow (\mu, B, \Gamma; \Sigma_{22}, \theta_2)$$

as given for example by Dawid *et al.* (1973, (A.1.2)), with $(1 - q) = (q + 1) - 2q$ in priors (5.36) and (5.29).

Remark 2. The posterior distribution of Theorem 5.2 is the predictive conditional distribution of ξ given X, Y, and Z that obtains from sampling from normal conditional distribution of X given Y with parameters $\mu_*(p \times 1), B_*(q \times p)$, and $\Gamma_*(p \times p)$ and prior proportional to

$$|\Gamma_*|^{-(q+1)/2}.$$

This is in accord with Remark 1 since the prior of the Theorem exactly recreates the joint predictive distribution of (ξ, Z) and it follows that the various routes to the conditional distribution cohere.

Remark 3 (Inference for replicates). Suppose $m > 1$, and each replicate corresponds to the same ξ. Then $\pi(\xi|X)$ as in Theorem 5.2 remains the natural exchangeable prior distribution. The predictive likelihood is given by eqn (5.34). Hence the posterior, $\pi(\xi|X, Y, Z)$, for ξ given X, Y, and Z, is the product of these two densities, and is proportional to

$$\{c_{m,n}(\xi)/c_{1,n}(\xi)\}^\nu / \{c_{m,n}^2(\xi) + (\bar{Z} - \hat{B}^T \xi)^T S_+^{-1}(\bar{Z} - \hat{B}^T \xi)\}^{(\nu+q)/2}. \quad (5.37)$$

The mean or mode of this posterior distribution also may be regarded as the natural generalization of the Krutchkoff estimator to $m > 1$ in the simple $p = q = 1$ case. See also the similar profile likelihood based function (2.27) and following discussion. Such estimators are consistent as both m and n tend to infinity.

The usual design requirements for accurate estimation of a regression relationship together with the need for model validation would suggest that the design of X in the calibration experiment should span the full range of likely x-values and be spread evenly enough to validate the model. If this

has been done then an exchangeable normal prior distribution as indicated
by Theorem 5.2 may not be too far off the mark. In such circumstances,
reversing the roles of X and Y and utilizing the predictive distribution of
X given Y is sensible from a Bayesian perspective, even though in model
(5.1) X is fixed by design.

5.7.1 INFORMATIVE PRIOR KNOWLEDGE

Other types of prior information may be introduced and applied using
Theorem 5.1. Informative forms of prior information utilizing a time series structure of random explanatory variables are discussed in Chapter
6. That chapter is directed at random calibration, and Bayes competitors
to the techniques of Chapter 4. Simple Bayes approaches to controlled
calibration that will work when $n - p - q - 1 \leq 0$ can be devised under
a Normal–Inverse–Wishart prior of Appendix B.1.1. A simple benchmark
implementation would have (a) vague prior knowledge for B, so that, in
Appendix A notation, $H_0^{-1} \to 0$, and (b) an $\mathcal{IW}(\delta; Q)$ prior distribution
for Γ with the scale matrix Q of a simple form, perhaps the intraclass correlation form of Dickey et $al.$ (1985). A full analysis stems from Theorem 5.1,
but the form of updating to the posterior $\mathcal{IW}(\delta + n; Q + S)$ for Γ suggests
an approximate estimate of ξ of

$$\hat{\xi}^* = (\hat{B}\tilde{\Gamma}^{-1}\hat{B}^T)^{-1}\hat{B}\tilde{\Gamma}^{-1}(Z - \hat{\mu}), \tag{5.38}$$

where $\tilde{\Gamma} = (Q + S)/(\delta + n - 2)$. If Q is of intraclass form there are
two hyperparameters in Q together with δ to specify, perhaps by cross-
validation. This may be compared with the usual estimator given in
eqn (5.15).

5.7.2 ON DISCARDING X-COMPONENTS

The inverse estimator $\check{\xi}(p \times 1)$ is constructed on a least-squares basis, one
component at a time. That is, to estimate a particular component of
ξ, none of the the other components of x are used. This estimator is
sensible in the controlled calibration situation with respect to a prior which
likens the ξ with past x, as proved in Theorem 5.2. In the controlled
calibration framework and in the absence of such prior information, there
is no definitive result which would point to ignoring the other components in
the construction of an estimator, and circumstances occur where this would
not be wise (see Sundberg (1982)). However, from practical experience the
suggestion of always specializing to $p = 1$ is not without merit. Intuitively it
does little good to collect other components if in future you are not to know
their value at the same time and on the same sample as the component of
interest. Specializing to $p = 1$ also offers considerable simplifications.

The next three sections of this chapter develop likelihood methods.

5.8   Profile likelihood with replications

In this section the profile likelihood for ξ is developed in controlled calibration and the normal error model, with n-independent training observations satisfying eqn (5.1), and m-independent prediction replicates from
eqn (5.10). The multivariate development exactly parallels the univariate
of Section 2.4. In this multivariate case though it will be seen that the
maximum likelihood estimator of ξ, the maximum of the profile likelihood,
is no longer a simple least-squares estimator linear in $\bar{Z}$ unless $p = q$. Because eqns (5.1) and (5.10) together constitute a curved exponential family
the MLE is highly non-linear. An explicit formula is given when the number of explanatory variables (p) is one. The profile likelihood allows you to
specify likelihood ratio confidence regions for the unknown p-vector ξ.

The unknown parameters of model of eqns (5.1), (5.10) are μ, B, Γ, and
ξ, where ξ is the parameter of interest and the others are nuisance parameters, whose removal is required for inference about ξ. One method of
elimination is by means of a prior distribution on the parameters followed
by integrating out the nuisance parameters in the posterior distribution
as in Section 5.7. As an alternative here, avoiding the need for specification of a prior distribution, the profile or maximum relative likelihood is
determined. As illustrated in Section 2.4 the method entails forming the
maximized likelihood *as if* ξ were known and normalizing so that the value
1 is attained when maximized over all the unknown parameters including ξ.
The profile likelihood is thus a function of ξ which has a maximum value 1
at a maximum likelihood estimator of ξ. The functional form gives a good
indication of the uncertainty to attach to this maximum likelihood estimator. By taking all values of ξ for which the profile likelihood is greater
than some fixed threshold, likelihood ratio confidence regions are provided.
The threshold may be specified by simple asymptotic chi-squared theory,
strictly here as the error covariance tends to zero. Modifications to this
simple profile likelihood idea to better accommodate uncertainty in the
plug-in parameters are discussed in the bibliography to Chapter 2.

For notational simplicity it is assumed that Y as well as X of eqn (5.1)
have been centred, *post hoc* in the case of Y. It has also been assumed that
Z has been adjusted to conform to the same origin as Y.

When ξ is assumed known, eqns (5.1) and (5.10) have the same unknown
parameters and they may be put together in a combined model:

$$Y_0 = 1_{n+m}\mu_0^T + X_0 B + \mathcal{E}_0, \qquad (5.39)$$

where Y_0 and $\mathcal{E}_0$ are $(n + m) \times q$ matrices, formed by augmenting Y and
$\mathcal{E}$ with Z and $\mathcal{E}^*$, respectively. Also $X_0 [(n + m) \times q]$ is the X matrix
augmented by m replicates of ξ and then centred by $-1_{n+m}\xi^T\{m/(n+m)\}$
so that

$$\mu_0 = \mu + B^T\xi\{m/(n + m)\},$$

which still involves a full complement of q independent unknowns even with ξ considered known. Note that X_0 is a function of ξ, the parameter of interest. Now the maximized log-likelihood for model (5.39) with independent multivariate normal errors satisfying eqn (5.2), is

$$\{(n+m)/2\}[q\{1 + \log(2\pi)\} - \log\{|\hat{\Gamma}(\xi)|\}], \qquad (5.40)$$

with

$$(n+m)\hat{\Gamma}(\xi) = (Y_0 - 1\hat{\mu}^T - X_0\hat{B}_0)^T(Y_0 - 1\hat{\mu}^T - X_0\hat{B}_0), \qquad (5.41)$$

the $(q \times q)$ matrix of sum of products of error from the least-squares fit, with the zero subscript on B serving as a reminder that this estimate depends on the temporarily assumed known ξ. Linear algebra enables eqn (5.41) to be written as an explicit function of ξ. By introducing the idempotent $(n+m)$ dimensional matrix (see Appendix D) given as

$$I - X_0(X_0^T X_0)^{-1} X_0^T$$

the right-hand side of eqn (5.41) may be written as

$$(Y_0 - 1\bar{Y}_0^T)^T \{I - X_0(X_0^T X_0)^{-1} X_0^T\}(Y_0 - 1\bar{Y}_0^T). \qquad (5.42)$$

Now from the facts,

$$(Y_0 - 1\bar{Y}_0^T)^T(Y_0 - 1\bar{Y}_0^T) = Y^T Y + Z^T Z - \frac{m^2}{m+n}\bar{Z}\bar{Z}^T$$

$$(Y_0 - 1\bar{Y}_0^T)^T X_0 = Y^T X + \frac{nm}{n+m}\bar{Z}\xi^T$$

$$X_0^T X_0 = X^T X + \frac{nm}{n+m}\xi\xi^T,$$

and the standard binomial inverse theorem as given in Appendix D, formula (5.42) becomes

$$Y^T Y + Z^T Z - m^2\bar{Z}\bar{Z}^T/(n+m)$$
$$+ \quad \{Y^T X + nm\bar{Z}\xi^T/(n+m)\}G\{\xi\xi^T/c^2(\xi) - G^{-1}\}G$$
$$\times \qquad \{Y^T X + nm\bar{Z}\xi^T/(n+m)\}^T \qquad (5.43)$$

where matrix G is given by eqn (5.5) and scalar $c^2(\xi)$ by eqn (5.6), dropping the m, n suffix for notational convenience. With residual sum of products S_+ defined by eqn (5.12), after some algebra eqn (5.43) becomes

$$S_+ + (\bar{Z} - \hat{B}^T\xi)(\bar{Z} - \hat{B}^T\xi)^T/c^2(\xi).$$

Since this is the right-hand side of eqn (5.41), taking determinants and using the determinantal identity, $|I + AB| = |I + BA|$ for conforming rectangular matrices,

$$|(n+m)\hat{\Gamma}(\xi)| = |S_+|\{1 + (\bar{Z} - \hat{B}^T\xi)^T S_+^{-1}(\bar{Z} - \hat{B}^T\xi)/c^2(\xi)\}. \qquad (5.44)$$

From eqn (5.40) the profile likelihood is

$$\{|\hat{\Gamma}(\xi)|/|\hat{\Gamma}(\tilde{\xi})|\}^{(n+m)/2}$$

where $\tilde{\xi}$ is a (or *the* if unique) maximum likelihood estimator, and reverting to the full notational form $c_{m,n}$ of eqn (5.6),

$$|\hat{\Gamma}(\xi)|^{\frac{n+m}{2}} \propto \{c_{m,n}^2(\xi)/[c_{m,n}^2(\xi) + (\bar{Z} - \hat{B}^T\xi)^T S_+^{-1}(\bar{Z} - \hat{B}^T\xi)]\}^{\frac{n+m}{2}}, \qquad (5.45)$$

so that the profile likelihood is proportional to the right-hand side of eqn (5.45).

Remark 1. The Bayes integrated likelihood (5.34) corresponding to a Jeffreys invariant vague prior is almost of the form (5.45): the difference being that the powers of the numerator and denominator of the Bayes posterior are $\nu/2$ and $(\nu + q)/2$ and not both $(\nu + p + q + 1)/2$ as in the profile likelihood.

Remark 2. The profile likelihood is also proportional to eqn (5.45) in the case of polynomial dependence of $E(Y)$ on X, that is when components of ξ are related.

Remark 3. if B were known and Γ unknown, the standard regression likelihood would avoid the form of dependence on ξ developed above. On the other hand if Γ were known the profile likelihood would be proportional to the exponential of $-\frac{n+m}{2}\text{trace}\{\Gamma^{-1}\hat{\Gamma}(\xi)\}$, thus also resulting in a dependence on $\hat{\Gamma}(\xi)$. Indeed the calibration problem in which data from model (5.1) provide an estimator $\hat{B}$ of B,

$$\hat{B} = B + E,$$

for use in eqn (5.10), has an errors in variables structure with the 'ratio' of error covariance matrices of E and $\mathcal{E}_*$ known. See Thomas (1991) for an exploitation of this errors in variables analogy. We prefer a more direct development of the maximum likelihood estimator following Brown and Sundberg (1987).

5.9 Maximum likelihood estimation

The profile likelihood in eqn (5.45) is proportional to

$$\{\nu + (R + ||\xi - \hat{\xi}||_K^2)/c_{m,n}^2(\xi)\}^{-(n+m)/2}, \tag{5.46}$$

with $\hat{\xi}$, K, and R given by eqns (5.15), (5.16), and (5.19), respectively. If $R = 0$, and this is necessarily so if $p = q$, then this is maximized by $\hat{\xi}$ the generalized least-squares estimator of ξ. Otherwise the maximum likelihood estimator $\tilde{\xi}$ is not equal to $\hat{\xi}$. It may be shown that a unique maximum likelihood estimator of ξ does exist under fairly general conditions and that the estimator is non-linear in $\bar{Z}$ and 'expands' the linear generalized least-squares estimator (see Brown and Sundberg (1987)). In the important special case $p = 1$ an explicit maximum likelihood estimator of the scalar ξ can be given as follows.

From formula (5.46) the profile likelihood is proportional to the $-(n + m)/2$ power of

$$\{\nu(n + m)/(nm) + f(\eta, R)\}, \tag{5.47}$$

with

$$f(\eta, R) = \{R + (\eta - \hat{\eta})^2\}/(1 + g\eta^2),$$

where

$$\eta = \xi\sqrt{K}, \quad \hat{\eta} = \hat{\xi}\sqrt{K}, \quad g = \left\{(1/n + 1/m)K\left(\sum_1^n x_i^2\right)\right\}^{-1}. \tag{5.48}$$

Maximization of the profile likelihood corresponds to function $f(\eta, R)$ minimimization. For the rest of this section the special case $\hat{\eta} = 0$ is excluded; it is easily dealt with separately. Consider equi-likelihood points $f(\eta, R) = c$. For any c this may be reorganized into the quadratic in η,

$$\eta^2(1 - cg) - 2\hat{\eta}\eta + (R - c + \hat{\eta}^2) = 0, \tag{5.49}$$

with 0, 1, or 2 roots. The case of a single root, '$b^2 = 4ac$' corresponds to a maximum or minimum and occurs when

$$4\hat{\eta}^2 = 4(R + \hat{\eta}^2 - c)(1 - cg).$$

This in turn is a quadratic in c,

$$c^2g - c\{(R + \hat{\eta}^2)g + 1\} + R = 0, \tag{5.50}$$

and may be solved for c. This has two real roots,

$$c = \{1 + g(\hat{\eta}^2 + R) \pm [\{1 + g(\hat{\eta}^2 + R)\}^2 - 4gR]^{1/2}\}/(2g). \tag{5.51}$$

For the sequel these two values of c are denoted by c_L and c_U, $c_L < c_U$. Since minimization of $f(\eta, R)$ corresponds to maximizing the likelihood, the values of η are

$$\tilde{\eta} = \hat{\eta}/(1 - c_L g), \quad \eta_0 = \hat{\eta}/(1 - c_U g), \tag{5.52}$$

where from now on $\tilde{\eta}$ denotes the particular value, $\tilde{\eta}(c_L)$, the maximum likelihood estimator of η.

For more insight into the positioning of the MLE, by reason of symmetry about centred x-values, without loss of generality, assume $\hat{\eta} > 0$. From eqn (5.52) and the sum and product of the roots of eqn (5.50) it follows that $\eta_0 \tilde{\eta} = -1/g$ so that η_0 and $\tilde{\eta}$ have opposite signs and consequently

$$0 \le c_L < 1/g < c_U.$$

Thus $\tilde{\eta}$ is an expansion of $\hat{\eta}$. Furthermore, $f(\eta, R) \to 1/g$ as $\eta \to \pm\infty$, from below at $+\infty$, from above at $-\infty$. Also eqn (5.49) may be reorganized by completing the square in η as the equation

$$(1 - cg)\{\eta - \hat{\eta}/(1 - cg)\}^2 = \hat{\eta}^2/(1 - cg) - \hat{\eta}^2 + c - R$$

with right-hand side an increasing function of c and a decreasing function of R. It follows that the root c_L is monotonically increasing as a function of R. At $R = 0$, $\tilde{\eta} = \hat{\eta}$ and hence $\tilde{\eta}$ goes from $\hat{\eta}$ to infinity as R goes from 0 to infinity. In fact for large R, from eqns (5.51) and (5.52),

$$c_L = (1/g)(1 - \hat{\eta}^2/R) + O(1/R^2),$$

and correspondingly

$$\tilde{\eta} = R/\hat{\eta} + O(1).$$

In summary the profile likelihood has a single minimum and a single maximum. The latter corresponds to the maximum likelihood estimate, given by $\tilde{\eta}$ of eqn (5.52) where c_L is the smaller root of eqn (5.51) and η relates to ξ by eqn (5.48). The maximum likelihood estimator $\tilde{\xi}$ is the same side of zero as $\hat{\xi}$; the minimum is the opposite side of zero. Suppose $\hat{\xi}$ is positive: the profile likelihood tends from above to a non-zero constant at infinite ξ; it tends to the same constant from below at minus infinity. Note that the value of the profile likelihood at $\xi = \pm\infty$ tends to zero only as $1/g \to 0$ and for this it is necessary that both n and m tend to infinity. The non-zero profile likelihood at infinity is an indication of the non-regular nature of the inference. The existence of a minimum on the opposite side of the origin to $\hat{\xi}$ implies that the profile likelihood will have an interval of highly unlikely values, to the right and left of which ξ-values are more likely.

It is natural to question whether the maximum likelihood estimator should be preferred to the simpler generalized least-squares estimator, $\hat{\xi}$. The answer must be in some doubt since the MLE offers an expansion of $\hat{\xi}$, and this is far less natural than the Bayes shrinkage of $\check{\xi}$ favoured by a

Bayesian analysis and prior which suggests ξ is like the x of calibration.
Also the form of MLE needs strictly to be recalculated for each new Z; it
is not simply linear in Z as is $\hat{\xi}$. However, if $\hat{\xi}$ is used, the influence of R
ought to be appreciated in the specification of uncertainty of the estimator.

5.10   Confidence intervals

Assessment of uncertainty of knowledge of ξ in a sampling framework may
be achieved by direct use of the profile likelihood function, giving regions
that include all ξ-values where the profile likelihood is greater than some
threshold. The threshold is then specified by exact, or failing that, asymptotic theory. These regions are usually called likelihood-based regions.

Explicitly, minus twice the log-likelihood ratio test statistic is, from
formula (5.46),

$$W(\xi) = (n + m)\{h(\xi, R) - h(\tilde{\xi}, R)\}$$

with

$$h(\xi, R) = \log\{\nu + (R + \|\xi - \hat{\xi}\|_K^2)/c_{m,n}^2(\xi)\}.$$

A likelihood-based confidence region for ξ is given by

$$\{\xi : W(\xi) \leq k\},\ k > 0. \tag{5.53}$$

Here k is specified by distribution theory so that the coverage is some
prespecified amount.

When $p = q$ an exact likelihood ratio test readily obtains. In this
case R is identically zero, $\tilde{\xi} = \hat{\xi}$, and $W(\xi)$ is a monotone function of the
left-hand side of eqn (5.14) with exact sampling distribution given by the
right-hand side F-distribution. As mentioned in Chapter 2 this gives a
direct interpretation of predictive Student confidence regions in terms of
the profile likelihood, including their pathological behaviour, which itself
is a direct consequence of existence of a minimum.

When $q > p$, as is typically the case, no exact likelihood ratio distribution theory is possible and asymptotic distribution theory is invoked.
Standard asymptotic distribution theory would take $k = k_p(\gamma)$, the upper
γ point of the chi-squared distribution on p degrees of freedom, to yield
an approximate $(1 - \gamma)100$ per cent confidence level. However, asymptotic
theory required for the limiting chi-squared distribution relies on increasing
information about ξ through repetitions, whereas here only n in the training data may be considered large. Consequently, since m is rarely large, the
limiting result needs to be justified by letting the error covariance matrix
tend to zero: small Γ asymptotics, as detailed for $p = 1$ by Brown and
Sundberg (1987). Rather than repeat their technical arguments here, the
form of the confidence interval is explored, and particularly the influence
of R.

Let us look at the confidence region of eqn (5.53) when $p = 1$ and (say) $\hat{\xi} > 0$. Transforming to η by eqn (5.48), from formula (5.47), $W(\eta)$ is a monotonic function of

$$\{l + f(\eta, R)\}/\{l + f(\tilde{\eta}, R)\},$$

with $l = \nu(m + n)/(mn)$, whose behaviour is determined by $f(\eta, R)$. The following facts have already been established:

1. $f(\tilde{\eta}, R) = c_L$, the minimum value of $f(\eta, R)$,
2. $f(\eta, R) \to 1/g$ as $|\eta| \to \infty$,
3. $c_L \to 1/g$ as $R \to \infty$.

Thus since there is only one minimum of $f(\eta, R)$ and consequently $W(\eta)$, at the MLE, and the minimum value tends to the value at infinity as R increases, it is evident that $W(\eta)$ gets flatter as R increases from zero. Hence likelihood ratio confidence intervals expand with R. What is more, region (5.53) is equivalent to

$$\{\eta : f(\eta, R) \le c^*(R)\},$$

where

$$c^*(R) = (l + c_L)[exp\{k/(m + n)\}] - l \stackrel{\text{def}}{=} a + bc_L$$

with $a > 0$, $b > 1$. For a bounded interval it is necessary and sufficient that $c^*(R) < 1/g$ and this will be aided by large n and γ and small R, since c_L is monotonically increasing in R. When $c^*(R) > 1/g$ the interval is unbounded, being typically doubly infinite with a centrally excluded set of values provided $c^* < c_U$, which increase monotonically with R. As an illustration using constructed values, for $R = 0, 4, 7, 20$, Fig. 5.1 plots $-W(\eta)$ and indicates likelihood ratio intervals for $p = 1$, $m = 1$, $q = 4$, $n = 50$, $\gamma = 0.05$, $g = 0.1$, $\hat{\eta} = 1$. For R larger than about 8 in this case $c^*(R) > 1/g$ and the intervals are unbounded. Since R is approximately chi-squared on $(q - p) = 3$ degrees of freedom, such discordant values of $R > 8$ can be expected about 5 per cent of the time. More generally this will depend on the signal-to-noise ratio and other terms that contribute to $1/g$. The intervals are respectively: $(-0.98, 4.10)$, $(-1.02, 8.51)$, $(-0.85, 71.58)$, and $(-\infty, -8.4) \cup (2.6, +\infty)$. Notice that in the last case of $R = 20$ the generalized least-squares estimator is not even included in the interval! The message is probably not that you ought to use the MLE; rather, it is that inference is very uncertain and the prediction experiment should be repeated and both prediction model and calibration model checked for concordance.

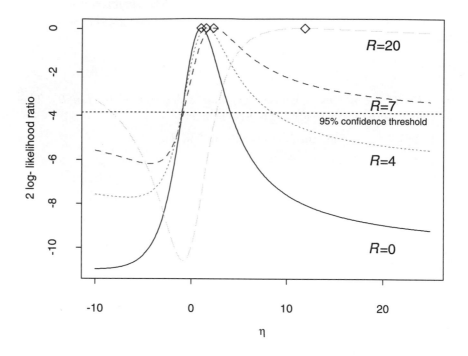

Fig. 5.1.   Plot of twice log-profile likelihood against η for four selected R, the MLE for each case is indicated by $\diamond$ and asymptotic 95 per cent confidence intervals consist of all η values for which each curve is above $-(1.96)^2 = -3.84$

5.11   Prediction diagnostics

Previous sections of this chapter have pinpointed R, defined by eqn (5.19), as a diagnostic in multivariate calibration. It has a known asymptotic distribution for $q > p$, and, when large relative to this distribution, indicates increasing uncertainty in one's estimate of the p-vector ξ, be it from a Bayesian or likelihood perspective. It has the properties of an asymptotic ancillary for ξ on $(q - p)$ degrees of freedom. In this section two further diagnostics are defined and a simple relationship between them described. They may be used to suggest outliers. A large value of any one of them may be indicative of anomalous behaviour.

The diagnostics assume the calibration and prediction models, (5.1), (5.2), (5.10), to be true. In practice, after calibration, many different samples are analysed corresponding to different and unknown ξ values. Consistent repeated departures from expected values lead us to question the model rather than identify outliers. However, since the ξ are unknown, whilst detection of changes from the calibration model are possible, correction of the regression model is not possible without further strong assumptions. The following development follows that of Brown and Sundberg (1989).

5.11.1 PREDICTION OUTLIERS

For this section, assume there are no prediction replicates ($m = 1$), and that the parameters, μ, B, and Γ are precisely estimated by $\hat{\mu}, \hat{B}$, and $\hat{\Gamma}$ so that they may be regarded as known. Under this assumption the prediction model (5.10) is solely of interest and ξ is the only unknown parameter. It may be written

$$Z = \hat{\mu} + \hat{B}^T \xi + \epsilon.$$

In fact premultiplying this by $\hat{\Gamma}^{-1/2}$ leads to a homoscedastic regression problem with model matrix $\hat{\Gamma}^{-1/2}\hat{B}^T$. The assumption of no information on ξ suggests the least-squares predictor of the response q-vector as

$$\hat{Z} = \hat{\mu} + \hat{B}^T \hat{\xi};$$

whereas the the assumption that

$$\xi \sim N_p(0, \Sigma),$$

where Σ as in eqn (5.24) is the covariance matrix of the x-vectors from the training data, leads to the inverse predictor,

$$\check{Z} = \hat{\mu} + \hat{B}^T \check{\xi}.$$

Now the prediction covariance matrix is

$$E(Z - \check{Z})(Z - \check{Z})^T = \hat{\Gamma}(\hat{\Gamma} + \hat{B}^T \Sigma \hat{B})^{-1}\hat{\Gamma} \stackrel{\text{def}}{=} \Theta^{-1}. \qquad (5.54)$$

This may be derived from Section 5.5 or directly from conditioning the partitioned $q + p$ random variate (Z^T, X^T). Suppose this is multivariate normal with covariance matrix

$$\begin{pmatrix} C_{zz} & C_{zx} \\ C_{xz} & C_{xx} \end{pmatrix},$$

where $\Sigma = C_{xx}$, and zero mean (corresponding to $\hat{\mu} = 0$ above). The regression mean of Z on $(X = x)$ is $B^T = C_{zx}C_{xx}^{-1}x$ with covariance matrix $\Gamma = C_{zz.x} = C_{zz} - C_{zx}C_{xx}^{-1}C_{xz}$. The regression mean of X on $(Z = z)$ is $\check{\xi} = C_{xz}C_{zz}^{-1}z$ and hence $\check{Z} = B^T\check{\xi} = B^T C_{xz}C_{zz}^{-1}z = Az$, defining A. Thus $I - A = \Gamma C_{zz}^{-1}$ and

$$E(Z - \check{Z})(Z - \check{Z})^T = E(I - A)ZZ^T(I - A)^T$$

$$
\begin{aligned}
&= \; \Gamma C_{zz}^{-1} C_{zz} C_{zz}^{-1} \Gamma \\
&= \; \Gamma C_{zz}^{-1} \Gamma \\
&= \; \Gamma (\Gamma + B^T \Sigma B)^{-1} \Gamma,
\end{aligned}
$$

proving eqn (5.54), since the estimates from calibration are assumed accurate.

With $x^T A x = ||x||_A^2$, defining

$$
R_X = ||\hat{Z} - \check{Z}||_\Theta^2, \tag{5.55}
$$

and

$$
R_B = ||Z - \check{Z}||_\Theta^2,
$$

Næs and Martens (1987) show that

$$
R_B = R + R_X.
$$

Under normality R_X and R are independent chi-squared random variables with p and $(q - p)$ degrees of freedom, and the result may be proved as an application of Cochran's theorem for quadratic forms. It may be noted that

$$
R = ||Z - \hat{Z}||_{\hat{\Gamma}^{-1}}^2 = ||Z - \hat{Z}||_\Theta^2,
$$

so that all three quadratics may be considered as having the same norming matrix Θ. The diagnostic R_X is likely to be large if the predictions from the two regressions differ greatly and this will be the case if ξ is extreme in composition (x) space. We may note in contrast that if ξ is at the (zero) mean of the x-values then there is no difference between the two predictions. A large value of R_X would suggest that particular care should be exercised in using $\check{\xi}$, the natural case predictor, since it shrinks towards the mean of the x-values.

The above assumes accurate estimates of parameters from the calibration. When this is not so, as when $n < q$, then the generalized least-squares estimator would need to be defined via generalized inverses. However, in near singular cases, one is far from the accurately estimated model required of the chi-squared distribution theory, and experimentation with the present dataset suggests that the measures R and R_X may be very inflated when there are not sufficient degrees of freedom in the training data to adequately estimate Γ. See Fujikoshi and Nishii (1984, Theorem 1) for one result when $n > p + q + 1$ with regression matrix B accurately determined but not necessarily Γ. When the number of training set observations is less than or close to $p + q + 1$ then the need for the incorporation of prior information about Γ may be irresistible, as in the informative Bayes estimator (5.38).

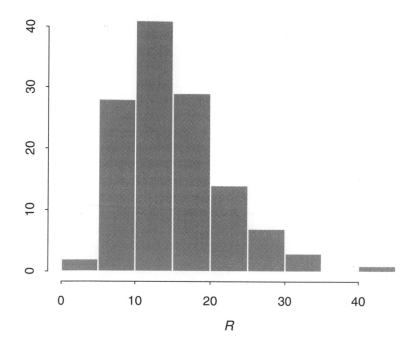

Fig. 5.2.   Sugar data, histogram of diagnostic R for 125 sample training data

The diagnostics are illustrated below on the sugars data of Example 1.6. Eighteen selected response wavelengths and the 125 observation training data are used to provide accurate estimates of the regression parameters. with precise estimates of the regression parameters from the training data, R is chi-squared on $(q-p)$ degrees of freedom and independently R_X is chi-squared on p degrees of freedom. Figures 5.2 and 5.3 give histograms of R and R_X for each of the 125 training sample observations in turn separated off as spectra for prediction. They are roughly commensurate with the chi-squared on 15 and 3 degrees of freedom. There are no particularly anomalous values.

The 21 values of R for the prediction data set are an order of magnitude larger. They are plotted in Fig. 5.4 against the squared prediction error averaged over the three components, computed from knowledge of their true composition. As an aside we may report that this average prediction error has by far the largest errors for glucose. It is here that further work might consider adding more wavelengths. Figure 5.4 shows an evident if modest correlation, dominated by exceptionally high error and R for the sixth observation. This observation is in fact the extreme composition of 25 per cent on each of the three sugars (see Table 1.5). The values of R_X

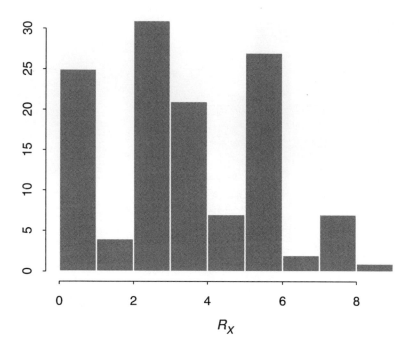

Fig. 5.3. Sugar data, histogram of diagnostic R_X for 125 sample training data

in Fig. 5.5 show a lesser correlation aside from intruding dominance again of the sixth observation. The first seven of the 21 prediction data spectra all show high values of R_X. They involve triple combinations of extreme compositions of 0 or 25 per cent for each of the three sugars.

It might be argued that trying to predict values of composition outside the data used in calibration breaks the cardinal rule of not extrapolating. However, it is never known for sure that future values will not be more extreme than the training data. What is more, the diagnostics serve to highlight when this might be a problem. Even though it is identified as a potential problem it is often still possible to predict composition of the sugars quite accurately, except in one or two cases. As a monitoring technique indeed it would seem to throw up far more suspicious cases than is really warranted. This seems inevitable since an atypical spectrum relative to the training data is all that is highlighted. We did separately use a 25 observation random subset of the 125 observation set as validation data, with the remaining 100 observations as training data and then the 25 values of R were much more typical of the reference chi-squared distribution (all but 2 of the values were less than 40 and the largest was 70). Values of R and R_X for the 21 observation validation set are so large that one would probably

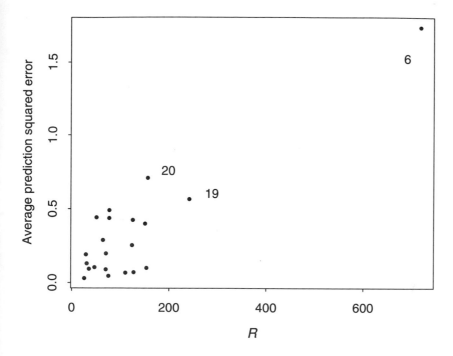

Fig. 5.4.   Sugar data, squared prediction error averaged over 3 sugars by diagnostic R

highlight three quarters of them as atypical. It may be suspected that the linear model breaks down for the prediction, although this is not borne out in further analysis in Chapter 7, at least for the selected wavelengths.

5.11.2   PREDICTION MODEL CHANGE

When it comes to examination of a sequence of Z values each one corresponding to a different value of ξ, then the focus shifts from individual outliers to systematic departures from the calibration model. Suppose there are t observations, with parameters precisely estimated as above, in the model

$$Z_i = \hat{\mu} + \hat{B}^T \xi_i + \epsilon_i \quad i = 1, ..., t. \tag{5.56}$$

Unfortunately but inevitably, since the left-hand side of eqn (5.56) only is observed, it is not possible to correct for systematic departures, without further strong assumptions. In prediction, a change in the distribution of Z from that of Y in the calibration sample, as measured by the prediction sample, $Z_1, \ldots, Z_t$ could result from a change in

1. (μ, B),

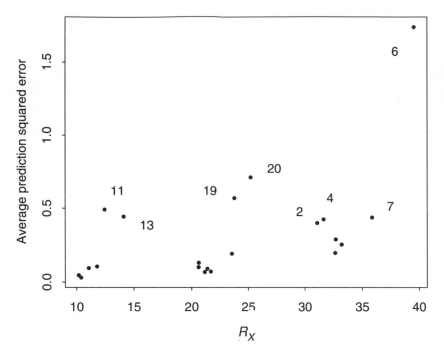

Fig. 5.5. Sugar data, squared prediction error averaged over 3 sugars by diagnostic R_X

2. Γ,

3. the distribution of ξ.

Any one or more of these may occur and from observations on just $Z_i, i = 1, \ldots, t$ it is impossible to disentangle which have occurred. In practice it seems to be not unreasonable to assume (1) has occurred and not (2) or (3). Within (1) one may wish to see whether change is in both μ and B or μ alone. The hypothesis of change in B alone is untenable from marginality considerations. Note that if (3) remains unchanged, then the $N_p(0, \Sigma)$ distribution of ξ has Σ known from the calibration data.

    It is clear from the canonical variates analysis of Section 5.5 that there are just p linear relations linking Y to X. Therefore to avoid investigating irrelevant change it is best to reduce model (5.56) canonically. The following preliminary reparametrization reduces the covariance matrices of both error and ξ to the identity. To reconstruct the untransformed notation from the sequel, all subsequent formulae must be transformed back by substitutions:

$$\mu \to \Gamma^{-1/2}\mu, \ B^T \to \Gamma^{-1/2}B^T\Sigma^{1/2}, \ Z_i \to \Gamma^{-1/2}Z_i. \qquad (5.57)$$

The parameters μ, B, and Γ are precisely estimated by least-squares and at these estimates you may choose orthogonal matrices, $Q(q \times q)$ and $P(p \times p)$ so that the transformed variables

$$Z_i^* = QZ_i, \; \xi_i^* = P\xi_i,$$

may be used to express estimated model (5.56) in the canonical form,

$$
\begin{aligned}
Z_i^{*(1)} &= \hat{\mu}_*^{(1)} + \hat{B}_*^T \xi_i^* + \epsilon_i^{(1)} & (5.58) \\
Z_i^{*(2)} &= \hat{\mu}_*^{(2)} + \epsilon_i^{(2)}, & (5.59)
\end{aligned}
$$

where $Z^{*(1)}$ and $Z^{*(2)}$ are p and $(q-p)$ dimensional vectors respectively, $\hat{\mu}_*$ is a new constant vector, $\hat{B}_* = \text{Diag}(\hat{\beta}_1, \ldots, \hat{\beta}_p)$ is a diagonal $(p \times p)$ matrix with singular values of $\hat{B}$ as diagonal elements, that is $\hat{\beta}_j^2, j = 1, \ldots, p$ are eigenvalues of $\hat{B}\hat{B}^T$, and $\epsilon_i^{(1)}(p \times 1)$, $\epsilon_i^{(2)}[(q-p) \times 1]$ are independent standard normal vectors. In fact the diagnostic R is completely specified by eqn (5.59), namely

$$R_i = ||Z_i^{*(2)} - \hat{\mu}_*^{(2)}||^2,$$

is the diagnostic for the ith observation, and is ancillary to ξ in this asymptotic formulation.

Now the orthogonal matrix Q is a function of $\hat{B}$ and transforms the estimated model to canonical form. If (μ, B) change from $(\hat{\mu}, \hat{B})$, so will both $\hat{\mu}_*$ and $\hat{B}_*$ in eqns (5.58) and (5.59), the latter from zero in eqn (5.59). However, although eqns (5.58) and (5.59) both provide testable information about (μ, B), only eqn (5.58) involves ξ as used for prediction so that change in (μ, B) manifest in eqn (5.59) will not affect the behaviour of $\hat{\xi}$ or $\check{\xi}$ for prediction of ξ. Thus our null hypothesis is $(\mu_*^{(1)}, B_*) = (\hat{\mu}_*^{(1)}, \hat{B}_*)$ versus the alternative $(\mu_*^{(1)}, B_*) \neq (\hat{\mu}_*^{(1)}, \hat{B}_*)$.

Marginally, averaging over the unobserved ξ_i^* distributed as $N_p(0, I)$, eqn (5.58) becomes

$$Z_i^{*(1)} = \mu_*^{(1)} + \delta_i^*, \tag{5.60}$$

where δ_i^* are independent identically $N_p(0, \Delta)$ with

$$\Delta = B_*^T B_* + I_p, \tag{5.61}$$

and change in mean can be identified by change in $\mu_*^{(1)}$ from $\hat{\mu}_*^{(1)}$ and change in B_* associated with Δ. Changes in B_* are only detectable in Δ up to a $(p \times p)$ orthogonal transformations as a left multiplier of B_*. Model (5.58) with ξ_i^* unknown is a factor analysis model, the aim here being to see whether mean μ or the factor loadings have changed. The indeterminacy in estimation of the loadings is a standard feature of such models, as analysed through the induced covariance matrix (5.61).

If the null hypothesis were $\mu_*^{(1)} = \hat{\mu}_*^{(1)}$ versus $\mu_*^{(1)} \neq \hat{\mu}_*^{(1)}$, with B_* constant throughout then the test could be accomplished through (5.60) merely by testing for a mean change with a known covariance matrix of the error. An estimate of change and confidence region also readily obtain.

With respect to possible change in both $\mu_*^{(1)}$ and B_*, a likelihood ratio test may be used. This entails comparing the log-likelihood under the null hypothesis and log-likelihood maximized over the alternative. Under the null hypothesis the log-likelihood is

$$- pt \log(2\pi) - (t/2) \left\{ \log |\hat{\Delta}| + (1/t) \sum_{i=1}^{t} ||Z_i^{*(1)} - \hat{\mu}_*^{(1)}||^2_{\hat{\Delta}^{-1}} \right\}, \quad (5.62)$$

where $\hat{\Delta}$ is given by eqn (5.61) with B_* set at its null value $\hat{B}_*$. Now, taking the full log-likelihood and maximizing first over $\mu_*^{(1)}$ gives

$$-pt \log(2\pi) - (t/2) \left\{ \log |\Delta| + (1/t) \sum_{i=1}^{t} ||Z_i^{*(1)} - \bar{Z}^{*(1)}||^2_{\Delta_*^{-1}} \right\}$$

and the remaining maximization over B_* is equivalent to minimization of

$$h(\Delta, V) = \log |\Delta| + \text{trace}(\Delta^{-1} V)$$

with

$$V = \sum_{i=1}^{t} (Z_i^{*(1)} - \bar{Z}^{*(1)})(Z_i^{*(1)} - \bar{Z}^{*(1)})^T / t$$

the sample covariance matrix of the t new p-vectors $Z_i^{*(1)}$. With Δ specified by eqn (5.61), the minimization is for $t > p$ given by the maximum likelihood result of Jöreskög used in factor analysis. Denoting $\phi_1, \ldots, \phi_p \geq 0$ the eigenvalues of the $p \times p$ covariance matrix V and setting

$$a_i = \max(\phi_i - 1, 0), \ i = 1, \ldots, p, \quad (5.63)$$

the minimum of $h((\Delta, V)$ is at $\Delta = \tilde{\Delta}$ with

$$h(\tilde{\Delta}, V) = \sum_{i=1}^{p} \{\log(a_i + 1) - a_i \phi_i / (a_i + 1) + \text{trace}(V)\};$$

see, for instance, Mardia *et al.* (1979, p. 265). The maximized log-likelihood to be compared with formula (5.62) is then

$$- pt \log(2\pi) - (t/2)h(\tilde{\Delta}, V). \qquad (5.64)$$

The reference value of unity in eqn (5.63) stems from the transformation to identity error covariance matrix in eqn (5.57). Twice formula (5.62) subtracted from formula (5.64) gives a log-likelihood ratio test which is asymptotically distributed as chi-squared with

$$p(p+1) - (p/2)(p-1) = (p/2)(p+3)$$

degrees of freedom, where the calculation of degrees of freedom takes into account the indeterminacy in $\hat{\Delta}_*$ up to a $p \times p$ orthogonal matrix.

Note finally the crucial assumption of a fixed known ξ distribution. Actually a likelihood ratio test for detection of a change in mean and covariance matrix of ξ under the assumption of (μ, B) fixed has precisely the same test statistic as that to detect change in $(\mu_*^{(1)}, B_*)$ under fixed ξ distribution. Since in providing likelihood ratio tests estimates of change are also specified, it is possible to correct for detected changes under this strong assumption. If the assumption of known fixed ξ distribution is deemed too strong then a detected change will need to be corrected by recalibration. A further warning note: normality has been assumed throughout and tests of variance are notoriously sensitive to normality assumptions (see Box (1953)) and such tests could be made more robust. Alternatively, and perhaps more usefully, you might derive graphical monitoring procedures of components of $Z_i^{*(1)}$ for $i = 1, \ldots, t$, looking for change in mean level or spread as observations accumulate.

5.12 Reprise

This chapter has considered multivariate regression and calibration from classical, likelihood, and Bayesian viewpoints. The focus has been controlled calibration, although the random calibration estimator was specified in eqn (5.20), to enable comparison with the controlled estimator (5.15) and to show how it can be derived as a Bayes estimator under controlled calibration under plausible prior assumptions. An example of a Bayes estimator for a more informative prior distribution is given as estimator (5.38).

The controlled calibration linear model gives rise to difficulties additional to those encountered in Chapter 2. Fundamentally this is because the model (5.1) with (5.10) represents a curved exponential family when $q > p$ (see for example Efron (1978) for a discussion of curved exponential families). The mean of the q-vector Z_i, the ith observation in eqn (5.10) is $\gamma = \mu + \xi^T B$ which has only p unknown additional parameters in its structure.

By virtue of Beer's law, chemists often use the multivariate controlled calibration model except that all estimators are as if there are no non-zero error covariances and $\Gamma = \sigma^2 I$. Often the X matrix of absorbances is denoted as $K^T(n \times q)$ leading to the description 'the K-matrix method' with estimator $(KK^T)^{-1}Kz$ of the p unknown compositions corresponding to the $(q \times 1)$ spectrum z. Such a procedure may be inefficient if, as experience suggests, the covariance is far from being proportional to the identity matrix. However, when there are only a modest number of observations relative to variables there is a trade-off with extra variability introduced by an estimated Γ in eqn (5.15). The simple hybrid Bayes estimator (5.38) offers an alternative way of coping with this. If the prior covariance matrix Q in the Bayes estimator were $\tau^2 I$ then the K-matrix method results as a limiting case when the degrees of freedom on the prior expectation becomes large.

A structural virtue of the incorporation of the full unknown $(q \times q)$ covariance matrix Γ, aside from its validity in practice, is the resultant link with the regression of X on y, as discussed in Section 5.5. It may also be noted that the estimators are in identical subspaces in certain singular cases, a result of Sundberg and Brown (1989).

Finally the mathematics of prediction of p compositions simultaneously is given. For a variety of reasons there is, however, considerable merit for estimation in specializing to $p = 1$ in the formulae and then carrying out p separate estimation steps.

6
Regression on curves

6.1 Introduction

This chapter extends the discussion of regression modelling to the situation where the set of explanatory variables is indexed by a variate, frequency, or time, and when plotted as in Fig. 1.5 each set of explanatory variables forms a discretized curve. These variables essentially form some sub-sampled sequence of a continuous random process, so that there is an order to the variables and often the sub-sampling will be at equally spaced intervals of the underlying continuous variate (frequency or time). The number, q, typically here large, of explanatory random variables, will be denoted $Y_{(q)} = (Y_1, Y_2, ..., Y_q,)$ and will be used to predict the response or interest variable, here univariate and denoted as X.

Training data are available for which both the interest variable and the explanatory variables are observed on n specimens. In this chapter we assume that the data may be viewed as obtained at random so that we are in the *random* or *natural* calibration case as opposed to the more problematic *controlled* calibration where the interest variable has been strictly designed in the training data. In effect the calibration paradigm just reduces to standard regression prediction except that we have reversed the roles of X and Y. As in Chapter 4, the number q of explanatory variables may be very much larger than the number of observations, and there is a high degree of indeterminacy in fitting linear models. The detergent example described in Section 6.9 has $n=12$ and $q = 43$ extracted frequencies, but had the full set of frequencies been used q would be 1168, leading to the full set of 12 $Y_{(q)}$ shown in Fig. 1.5. In this example component 2 of Table 1.4 serves as the twelve X-observations for calibration.

The methodology is Bayesian and a variety of models stipulating a wide range of differing assumptions are discussed. With the high degree of indeterminacy, a well-structured prior distribution appropriate to the sequenced nature of the data becomes essential. Techniques discussed earlier in Chapter 4 involve principal components regression, ridge regression, continuum regression, and partial least-squares. None of these approaches takes account of the contiguity of regressors, the focus of this chapter. This approach would seem to be applicable more generally to re-

peated measurements problems in medicine and social science (see Jones
and Ackerson (1990) for a stationary Gaussian process model in this context). Applications may be as diverse as from chemometrics to measuring the Los Angeles Olympic marathon (see Smith and Corbett (1987)).
Mostly the description will follow chemometric applications. The methodology has both longitudinal data and time series constituents. The chapter
deals primarily with a class of prior distributions proposed by Brown and
Mäkeläinen (1992). Some alternative prior specifications are reviewed in
Section 6.10. The chapter ends with a sketch of an extension to spatial
prior distributions, a methodology which provides a Bayesian alternative
to kriging (Matheron, 1971) in geostatistics.

6.2   The model

We shall denote the data matrix, which is of order $n \times (q+1)$, by

$$Y^q = (X, Y_{(q)}). \tag{6.1}$$

Here X is the vector containing the observations on the regressand and
$Y_{(q)}$ is the matrix of regressors. One might like to think of X as the zeroth
variable, corresponding to its initial position in the $q + 1$ variables. We
shall consider random regressions and make the assumption that Y^q is a
sample from a $(q+1)$-variate normal distribution having mean zero, covariance matrix Σ. The assumption of zero mean involves no loss of generality
provided the mean vector is assumed known and we will relax this assumption when necessary later. In view of the supposition that regression, that
is the conditional distribution of X given $Y_{(q)}$, is the centre of interest,
this modelling of the joint distribution amounts to making more extensive
assumptions than are strictly necessary. They are needed to enable us to
specify prior assumptions coherently over submodels and envisaged refined
data as discussed in Section 6.4.

A future observation

$$Z^q = (\xi, Z_{(q)}) \tag{6.2}$$

is envisaged from the same distribution but conditionally independent of
data (6.1) given the parameters of the distribution. Finally, consider the
prior specification of the unknown covariance matrix Σ. Its distribution is
taken to follow an Inverse Wishart distribution with positive definite scale
matrix M^q.

To facilitate Bayesian manipulations and to enable extension to count-
ably infinite dimensional distributions, we use the notation introduced by
Dawid (1981) and given in Appendix A. In summary,

$$
\begin{align}
Y^q \mid \Sigma \quad &\sim \quad \mathcal{N}_{n,q+1}(I_n, \Sigma), \tag{6.3}\\
\Sigma \quad &\sim \quad IW_{q+1}(\delta, M^q), \qquad \delta > 0, M^q > 0, \tag{6.4}
\end{align}
$$

$$Z^q \mid \Sigma \quad \sim \quad \mathcal{N}_{1,q+1}(1, \Sigma), \qquad\qquad (6.5)$$

$$Y^q \text{ and } Z^q \quad \text{are} \quad \text{independent, given } \Sigma. \qquad\qquad (6.6)$$

Here given Σ the rows are independent and within any row the covariance matrix is Σ. More generally if a random matrix is $\mathcal{N}(A, B)$ then $a_{ii}B$ and $b_{jj}A$ are the covariance matrices of the ith row and jth column, respectively. The shape parameter δ above implies $\delta + (q + 1) - 1$ degrees of freedom in the more standard notation, and avoids immediate notational difficulties when q is infinite. In some contexts it may be natural to consider an infinite sequence of regressors, when each of Y^q, Z^q, and M^q are infinite with the understanding that for any finite subset appropriate conforming submatrices are chosen.

The above model has well-known deficiencies in the range of beliefs which are expressible by varying the hyperparameters, scalar δ and $(q + 1) \times (q + 1)$ prior scale matrix M^q. Only one parameter δ is available with which to determine the precision of specification of M^q. One way around this is through a *generalized Inverse Wishart* as developed by Brown *et al.* (1993a) which utilizes the Bartlett decomposition and effectively allows the degrees of freedom to vary over partitions of the scale matrix, whilst retaining analytic tractability.

A different criticism has been levelled by Dawid (1988), who has elucidated a *determinism* of the prior in prediction in the infinite-dimensional case. With probability 1 it is possible to predict the response X exactly from the prior alone. It is not of course clear in any finite setting which subset of the responses will achieve perfect prediction.

Recognizing the attractive simplicity of the Normal–Inverse–Wishart, we have adopted it, warts and all, watchful of unnatural features emerging from the posterior distribution. A similar grudging acceptance is shown in similar circumstances by Lindley (1978, Section 6), further reported by Dickey *et al.* (1985). Whereas Dickey *et al.* (1985) adopt the rather special intraclass prior expected covariance, Mäkeläinen and Brown (1988) develop a class of prior expected covariance matrices which accept an ordering of the predictive variables. In the present chapter, the prior expected covariance matrices form a rich class, with prior structures for M^q described in Section 6.5. It is as if Y^q is sub-sampled from an underlying continuous parameter Gaussian process. We have, however, avoided directly assigning a prior for a Gaussian process. Rather the structure is in the secondary hyperparameter level. This has an added bonus. We offer a more robust analysis by avoiding the strong assumption of a class of stationary Gaussian processes and relegating such structures to the hyperparameters of the Inverse Wishart prior; see Chen (1979) for a similar attitude to imposing structural information.

The more usual approach to regression prediction works with the conditional distribution of the response given the predictive variables. By working less directly simply with the full joint distribution we are able to incorporate coherence across assignments of prior information for varying numbers of regressors (see Section 6.4).

6.3   The posterior distribution

Future prediction envisages estimating ξ from $Z_{(q)}$ given the information about Σ as updated through observing Y^q and $Z_{(q)}$. We repeat the argument of Mäkeläinen and Brown (1987). The conditional distribution of ξ given Y^q and $Z_{(q)}$ is found by two consecutive conditionings in a matrix-variate T-distribution. To formulate the result needed suppose $T \sim \mathcal{T}_{r,s}(\delta; H, K)$. Here this matrix-T distribution denotes the random matrix which given Σ is $\mathcal{N}(H, \Sigma)$ and Σ is $IW(\delta, K)$. Consider the partitions $T = (T_1, T_2)$ and $K = [K_{ij}]$ with T_i an $r \times s_i$ matrix and K_{ij} an $s_i \times s_j$ matrix $(i, j = 1, 2)$ so that $s_1 + s_2 = s$. Assuming that K_{11} is positive definite we have (see Appendix A)

$$T_2 - T_1 B_{21}^T \mid T_1 \sim \mathcal{T}_{r,s_2}(\delta + s_1; H + T_1 K_{11}^{-1} T_1^T, K_{22\cdot1}), \qquad (6.7)$$

where $B_{21} = K_{21} K_{11}^{-1}$ and $K_{22\cdot1} = K_{22} - B_{21} K_{12}$. The formula indicates the linearity of the regressions in a T-distribution. Also note that $T^T \sim \mathcal{T}_{s,r}(\delta; K, H)$.

By the assumptions (6.3)–(6.6) and by the definition of a T-distribution we have that

$$\begin{bmatrix} Y^q \\ Z^q \end{bmatrix} \sim \mathcal{T}_{n+1, q+1}(\delta; I_{n+1}, M^q).$$

By the transpose of eqn (6.7),

$$Z^q \mid Y^q \sim \mathcal{T}_{1, q+1}(\delta + n; 1, N^q) \qquad (6.8)$$

where

$$N^q = M^q + (Y^q)^T Y^q. \qquad (6.9)$$

Thus N^q is the predictive covariance matrix of Z^q. We shall write

$$M^q = \begin{bmatrix} m_{00} & m_{0(q)} \\ m_{(q)0} & M_{(q)(q)} \end{bmatrix}$$

and similarly for N^q. Using (6.7) in (6.8) we finally find that

$$\xi \mid Z_{(q)}, Y^q \sim Z_{(q)}(b_{0(q)}^n)^T + \mathcal{T}_{1,1}(\delta^*; a, c) \qquad (6.10)$$

where

$$b_{0(q)}^n = n_{0(q)}(N_{(q)(q)})^{-1} \qquad (6.11)$$

and

$$a = 1 + Y_{(q)}N_{(q)(q)}^{-1}Y_{(q)}^T; \quad c = n_{00\cdot(q)}; \quad \delta^* = \delta + n + q.$$

In standard notation $T_{1,1}(\delta^*; a, c)$ above is a Student-t distribution on δ^* degrees of freedom scaled by $\sqrt{(ac/\delta^*)}$. Also notice that by the linearity of the regressions and since N^q is the covariance matrix of the predictive distribution of Z^q,

$$n_{00\cdot(q)} = n_{00} - b_{0(q)}^n n_{(q)0},$$

when scaled by appropriate degrees of freedom, is the residual variance in the regression of ξ on $Z_{(q)}$ in that distribution.

Equation (6.11) has a familiar form since by eqn (6.9)

$$N_{(q)(q)} = M_{(q)(q)} + Y_{(q)(q)})^T Y_{(q)(q)};$$

and from standard Normal Bayesian regression analysis (see for example eqn (7) of Lindley and Smith (1972)), eqn (6.11) is that estimate obtained from a prior covariance proportional to the *inverse* of $M_{(q)(q)}$, when the prior mean of the regression coefficients is eqn (6.11) with M^q in place of N^q. We will have cause to utilize this later. In the case of an intraclass correlation structure of Lindley (1978) it explains the negative prior correlation between regression coefficients noted by him when ρ is non-negative. Naively it reflects a scale property: scaling Y_j by a constant c scales the coefficient β_j of Y_j in the regression of X on $Y_{(q)}$ by $1/c$.

6.4   Coherent refinements

We are interested in the submatrices of the $n \times q$ regressor matrix

$$Y_{(q)} = (Y_1, Y_2, \ldots, Y_q)$$

such as

$$Y_{[q]} = (Y_1, Y_3, Y_5, \ldots, Y_q),$$

where q is odd, which obtain on thinning of the regressors. Conceptually at the same time we are interested in possible refinements of the data, whereby had initially only $q/2$ regressors $Y_{[q]}$ been observed we might wish to contemplate the prior distribution *had q* regressors $Y_{(q)}$ been taken. More generally we might think of $[q]$ as any subset of (q), the integers $1, 2, \ldots, q$.

The prior distribution for the regression coefficients changes in a complicated way as one drops or adds regressors. By assigning the prior distribution to the joint distribution of Y^q, as in Section 6.2, coherence of regression coefficient assignments is assured. If we had observed only observations at 4 nm intervals instead of 2 nm so that $q/2$ (q even) alternate variables are utilized, then the prior probability assignment for this subset would

1. correspond to the marginalized prior distribution,

2. be structurally generated by the same prior considerations which led to the generation of the prior for the refined set of variables.

Whereas (1) amounts to de Finetti's notion of coherence (see, for example, de Finetti (1974)), (2) provides stronger requirements and might be termed *structural* coherence. It has similarities with the extendibility notion in exchangeability. Readers may be reminded that such notions of coherence refer only to the prior distribution, the posterior distribution will certainly differ as a result of data at different resolutions, and should rightly do so.

A special feature of T-distributions is that for any subset $[q]$,

$$\xi \mid Z_{[q]}, Y^q \sim \xi \mid Z_{[q]}, Y^{[q]},$$

where $Y^{[q]} = (X, Y_{[q]})$. That is, if, at a future time when ξ is to be predicted by means of $Y_{(q)}$, the researcher should decide to use only $Z_{[q]}$, there would be no advantage in having retained the regressors in $Y_{[q]^c}$ where $[q]^c$ denotes the complementary even integers. This property, related to S-ancillarity, is discussed in Appendix B.

The posterior predictive distribution of ξ given the data $Y^{[q]}$ and $Z_{[q]}$ is then analogous to form (6.10) with $[q]$ replacing (q) and q replaced by $(q+1)/2$ or $q/2$ in the degrees of freedom depending on whether q is odd or even.

6.5   Contiguous prior structure

We will be concerned with the form of $M^q = (\delta - 2)E(\Sigma)$, where $E(\Sigma)$ is the prior expectation (which exists for $\delta > 2$) of the covariance matrix of the $q+1$ variables. The prior covariances of the regressand with the regressors, $m_{0(q)}$, may often be such that all q covariances are judged equal. At other times it may be that certain regressors or regions of regressors are more predictively important than others and in principle such beliefs may be incorporated. With this proviso, we usually take the covariances equal and make the common value zero.

The $q \times q$ matrix $M_{(q)(q)}$ is our major concern and is generally assumed to correspond to the covariance kernel of a sub-sampled stationary Gaussian random function. This involves a rich class of possibilities (see for example Yaglom (1987)). One subclass of these Gaussian random functions involves the kernel

$$\rho(\tau) = \exp(-\alpha \mid \tau \mid^\kappa), \alpha > 0$$

where $0 < \kappa \leq 2$, with an AR(1) embedded in $\kappa = 1$. Note though that in discrete time negative correlation is allowable but precluded in continuous time. Our methods apply to more general autoregressive processes, but with the above $\kappa = 1$ the simplest and perhaps most important special case. Suppose for example one has equal spacing with $\kappa = 1$. The form of covariance is an AR(1),

$$M_{(q)(q)} = \{k/(1-\rho^2)\} \begin{pmatrix} 1 & \rho & \rho^2 & \cdots & \rho^{q-1} \\ \rho & 1 & \rho & \cdots & \rho^{q-2} \\ \vdots & \vdots & \vdots & \vdots & \vdots \\ \rho^{q-1} & \rho^{q-2} & \cdots & \rho & 1 \end{pmatrix}. \qquad (6.12)$$

This has the same form as the prior covariance matrix $M_{[q][q]}$ for the $q/2$ thinned variables obtained by odd rows and columns of $M_{(q)(q)}$ except that the correlation parameter is now ρ^2, so that structural coherence is preserved. The same could not be said, for example, of a moving average structure for $M_{(q)(q)}$ when for a MA(1) the thinned covariance matrix would be proportional to an identity matrix and would not have the same structural form as for the q variables. Thus an MA(p) is precluded in modelling $Y_{(q)}$. Since the inverse of the covariance matrix of an MA(p) is the covariance matrix of an AR(p), we preclude an autoregressive structure for a prior distribution describing the regression coefficients in the conditional distribution. The continuous time parameter version of an MA(1) is further precluded by stationarity of the random process. Doob (1953, Chapter 11, Section 10) in treating the continuous analogue of an autoregressive-moving average process, an ARMA(p, s), defines the process in terms of a rational spectral density,

$$f(\omega) = \text{const} \frac{|(\omega-a_1)\ldots(\omega-a_s)|^2}{|(\omega-c_1)\ldots(\omega-c_p)|^2}.$$

For stationarity it is necessary and sufficient that the spectral density be non-negative and *integrable*. Hence it is necessary that the autoregressive order p be greater than the moving average order, s. (See also Yaglom (1987, Example 9, p. 133–36).) Except in the case of multiple roots of the spectral denominator, the covariance function is

$$\rho(\tau) = \sum_1^p \{C_j\cos(2\pi c'_j\tau) + D_j\sin(2\pi c'_j\tau)\}\exp(-2\pi c''_j\tau)$$

where C and D are constants not involving τ, and c'_j and c''_j are respectively the real and imaginary parts of c_j.

Thus an ARMA(2,1) would be allowable as would pure autoregressive structures for $M_{(q)(q)}$ of second and higher order since they also allow structural coherence. For example with equal spacing and an AR(2) process, the autocorrelation function might be

$$A\rho_1^\tau + B\rho_2^\tau$$

where ρ_1 and ρ_2 are roots (assumed unequal) of an auxiliary quadratic equation and stationarity imposes conditions on allowable ρ_1 and ρ_2. Thus it is evident that the thinned process will have the same form of autocorrelation function.

For a simple non-stationary dynamic model applied directly to the regression coefficients in the calibration of the Olympic marathon, see Smith and Corbett (1987); and for a smoothness prior similar to theirs see Polasek (1985).

6.6 Augmentation and computation

The posterior mean of the regression coefficients is given by eqn (6.11) and involves solving for the q-vector b the q equations given by

$$m_{(q)0} + Y_{(q)}^T X = (Y_{(q)}^T Y_{(q)} + M_{(q)(q)})\, b. \qquad (6.13)$$

When q may be of order one thousand this is a formidable task as it stands. If M is proportional to the identity matrix, as in ridge regression, we may orthogonally transform the problem and solve $\min\{q, n\}$ equations. This represents a substantial saving as in our applications n is very much smaller than q. This route is not open with our more general structure for M. However, an alternative route adapted from ridge regression as described in Section 4.4 not only enables us to speedily solve eqn (6.13) but also to avoid setting up this 'squared' equation in the first place.

Suppose we can analytically specify a $q \times q$ matrix A, preferably upper triangular, such that $A^T A = M_{(q)(q)}$. Now we augment the $n \times q$ data matrix $Y_{(q)}$ with q rows A, forming an $(n + q) \times q$ matrix $Y_{(q)}^*$. At the same time we augment the n-vector X by a q-vector $A^{-1} m_{(q)0}$, forming an $(n + q) \times 1$ response vector X^*. Note that this latter augmentation is trivial if $m_{(q)0}$ is zero, and is easy if A and hence A^{-1} is triangular. Standard regression algorithms can now be applied as if $Y_{(q)}^*$ is the design matrix and X^* the response vector. In particular one can apply the QR-decomposition and avoid 'squaring' quantities, simply back-solving a set of triangular equations; see for example Thisted (1988). With the augmentation matrix A already triangular one can effect the QR decomposition by a series of Givens elementary planar rotations of order q^2 in number, instead of the order q^3 needed to solve eqn (6.13) directly, with the added bonus of more numerical accuracy. Thisted rightly emphasizes the instability of the binomial inverse theorem (see Appendix D), often used to good effect for analytical data augmentation.

Lemma 6.1. *We can analytically construct an upper triangular $q \times q$ matrix square root A for any AR(p) sequence.*

We first demonstrate this for an AR(1). Referring to eqn (6.12), let $K = (1/k)M_{(q)(q)}$. The requisite matrix square root of K is C where $C^T C = K$ and

$$C = \begin{pmatrix} 1 & \rho & \rho^2 & \cdots & \rho^{q-2} & \rho^{q-1}/\sqrt{(1-\rho^2)} \\ 0 & 1 & \rho & \cdots & \rho^{q-3} & \rho^{q-2}/\sqrt{(1-\rho^2)} \\ 0 & 0 & 1 & \cdots & \rho^{q-4} & \rho^{q-3}/\sqrt{(1-\rho^2)} \\ \vdots & \vdots & \vdots & \vdots & \vdots & & \vdots \\ 0 & 0 & 0 & 0 & 0 & & 1/\sqrt{(1-\rho^2)} \end{pmatrix}$$

This is derived by noting that the inverse of K is tridiagonal with $-\rho$ off diagonal and $(1+\rho^2)$ on diagonal apart from the $(1,1)$ and (q,q) elements which are unity. The square root of this matrix is U where $U^T U = K^{-1}$ and

$$U = \begin{pmatrix} 1 & -\rho & 0 & \cdots & & 0 \\ 0 & 1 & -\rho & 0 & & \cdots \\ \vdots & \vdots & \vdots & \vdots & & \vdots \\ 0 & \cdots & \cdots & 1 & & -\rho \\ 0 & \cdots & \cdots & 0 & \sqrt{(1-\rho^2)} \end{pmatrix}.$$

This structure is inherent in the autoregression

$$Y_t = (1 - \rho L)^{-1} \epsilon_t$$

with L the lag operator and $\{\epsilon_t\}$ uncorrelated errors, augmented to q dimensions from the $q - 1$ differences. The U matrix for an AR(2) will have first $q - 2$ rows obtained from $1 - \phi_1 L - \phi_2 L^2$ and can be augmented by a further two rows so as to remain triangular and such that $U^T U = K^{-1}$. The inverse of U is then readily constructed and will be triangular.

Similar constructs to those above using Levinson's recursion may be found in Brockwell and Davis (1987), Chapters 5 and 8.

6.7   Mean modelling

The models above all assume that the mean vector of Y^q and also for Z^q is (i) known and (ii) constant from sample to sample. One may wish to relax these assumptions. When the mean vector is unknown it may be totally unstructured. Alternatively it may be assumed that $Z_{(q)}$ has a mean which is itself subsampled from some Gaussian process. If the mean is not constant from sample to sample then a hierarchical functional prior could be contemplated. This would be particularly apt in near infrared spectroscopy with ground solid preparations where variation in particle size serves to bodily shift the functional form (see Fearn (1983) for the adverse effect of this on routinely applied ridge regression, and the multivariate scatter corrections of Martens et al. (1983)).

However, for this chapter we will only explicitly consider a constant but unknown mean vector with an associated vague prior distribution, when posterior predictive distributions analogous to form (6.10) may be produced with deviations measured from the training sample mean vector and degrees of freedom as before or reduced by one, depending on the prior limit adopted. A full natural conjugate approach may be extracted from Ando and Kaufman (1965) or may be injected into Section 6.2 using Appendix B. The joint prior distribution of (Y^{qT}, Z^{qT}) is again matrix-variate Student and the required results found by two successive conditionings as before.

6.8 Hyperparameter estimation

The prior expected covariance matrix $M_{(q)(q)}$, as autoregressively structured in Section 6.5, has a number of unspecified hyperparameters. In the case of the AR(1) structure of eqn (6.12), there are two hyperparameters, k and ρ. We might envisage a further prior distribution on these hyperparameters. More pragmatically, following for example Lindley and Smith (1972), from the posterior Student distribution given in eqn (6.12), we may form modal posterior estimates of k and ρ. Alternatively, as in the example described in the next section, we choose these parameters by cross-validation. This is preferred as it is likely to be more robust to departures from the chosen simple Normal–Inverse–Wishart model.

6.9 Detergent example

The dataset consists of seven bands of frequencies chosen from 1168 mid-infrared equally spaced frequencies, as described in Example 1.5. The 43 frequencies were as chosen by a method described in Chapter 7 (a subset of the 47 selected there). In principle, however, one would keep all 1168 wavelengths but in practice there are good reasons, both computational and model based, for extracting this subset. Only twelve samples are available so that even with the wavelength selection a linear model is underdetermined.

The full data $Y_{(q)}$ are plotted as 12 samples of absorptions at the 1168 wavelengths in Fig. 1.5, and appear continuous to the resolution of printing. The twelve detergent mixtures contain 5 ingredients carefully designed as given in Table 1.4. We treat the data as randomly generated, however, and focus on one ingredient, the second component in Table 1.4. Figure 6.1 gives the same twelve absorption curves plotted against wavelength number with the mean spectrum subtracted from each spectrum and the 12 graphs ordered from the lowest value of the second component (6.99 per cent by weight) to the highest (13.22 per cent by weight). This exploratory plot may identify unusual or useful features in the data *vis a vis* relationship to the response of interest, component 2. It was originally suggested to me by Richard Lockhart. Here there does not appear to be any obvious trend. The five-ingredient data are further analysed, taking account of contiguity, using splines and autoregressive error structure in Denham and Brown (1993).

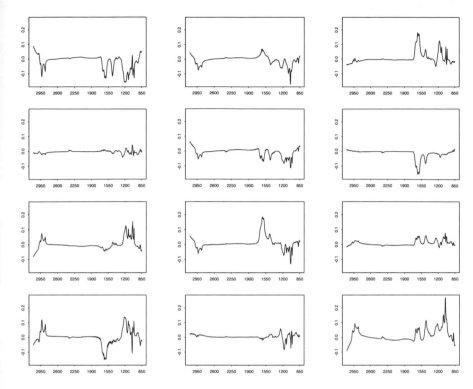

Fig. 6.1. Detergent data, 12 absorbance spectra, mean subtracted, ordered by component 2 composition, versus frequency (cm^{-1})

Table 6.1 gives cross-validated, that is leave-one-out, mean-squared prediction errors for the 12 observations at a grid of values of k and ρ. The minimum value 0.0309 at $k = 10^{-9}$ and $\rho = 0.7$ gives a root mean-squared prediction error of 0.18 and represents 99.5 per cent of the variation explained. It demonstrates the impressive accuracy of the infrared spectrometry. Statistically this is all the more reassuring given the $31 \ (= 43 - 12)$ degrees of indeterminacy in the implicit linear model. In this case this method of analysis has a 17 per cent smaller prediction mean-squared error than ridge regression, as embodied in the first row of Table 6.1 with $\rho = 0$. This is not quite the dramatic improvement which one might hope for in this example. Distributionally, the posterior predictive Student form of (6.10) may be used for prediction intervals.

6.10   Other prior assumptions

The autoregressive moving average covariance structure defined in Section 6.5 offers a flexible and wide choice of prior distribution, but may not

Table 6.1.   Detergent data, prediction mean-squared errors, cross-validated ($\times 10000$)

	k					
ρ	10^{-9}	10^{-7}	3×10^{-4}	7×10^{-3}	4×10^{-1}	2.7
0	405	405	407	420	374	546
0.50	322	322	327	379	356	3068
0.70	309	309	317	382	584	9791
0.90	363	363	371	415	489	3753

always be the most appropriate for the application in hand. In this section a few simple alternatives are proposed. Perhaps one of the simplest is the partially exchangeable covariance structure in which $m_{00} = k$, $m_{0(q)} = ka1$ and

$$M_{(q)(q)} = k \begin{pmatrix} 1 & \rho & \rho & \cdots & \rho \\ \rho & 1 & \rho & \cdots & \rho \\ \vdots & \vdots & \vdots & \vdots & \vdots \\ \rho & \rho & \cdots & \rho & 1 \end{pmatrix}. \tag{6.14}$$

This is given as eqn (5.1) of Mäkeläinen and Brown (1988). It specifies a common mutual correlation of ρ between the regressor (spectral) variables, and a common correlation a, usually zero between each of the regressor variables and the regressand. Indeed with $m_{(q)0} = 0$ the estimator of the regression coefficients, the solution b of eqn (6.13) with $M_{(q)(q)}$ given by eqn (6.14), is easy to conceive of as a generalized ridge estimator as follows. An intraclass correlation matrix such as eqn (6.14) may be diagonalized by the fixed orthogonal matrix Q (not depending on ρ and k), the first column (eigenvector) of Q proportional to the unit vector and the other $(q - 1)$ columns any spanning set of contrast vectors (see Appendix D). Correspondingly the first eigenvalue is $\lambda_1 = k[1 + (q - 1)\rho]$ and all the other eigenvalues are $k(1 - \rho)$. Thus if the model is written as

$$X = (Y_{(q)}Q)(Q^T \beta) + \epsilon$$

then with

$$Y^* = Y_{(q)}Q, \quad \beta^* = Q^T \beta, \quad b^* = Q^T b$$

eqn (6.13) becomes

$$Y^{*T} X = [Y^{*T} Y^{*T} + \text{Diag}(\lambda_1, \ldots, \lambda_q)] b^*.$$

This corresponds to a normal, mean zero, variance σ^2/λ_1 prior on $\beta_1^* = (\beta_1 + \ldots + \beta_q)/\sqrt{q}$; and independently normal, mean zero, variance $\sigma^2/[k(1-$

$\rho)]$ prior distributions for the other $(q-1)$ entries of β^*, that is the $(q-1)$ contrasts in the original regression coefficients β. Thus the sum of the coefficients is shrunk towards zero and the coefficients are induced to be similar. If k is large and ρ near 1 so that $k(1-\rho)$ is small then the sum of the coefficients will be shrunk to close to zero whereas the prior information will be otherwise uninformative. The precise nature of the shrinkage will be a compromise between the prior and data.

The estimator is further motivated by Fearn (1992) as the solution of a minimization problem. For fixed k and ρ the estimator, may be seen to be that β which minimizes $S+k\{\sum_1^q \beta_j^2+\rho(\sum_1^q \beta_j)^2\}$, where S is the residual sum of squares. As he notes, the shrinkage of the sum of regression coefficients is particularly apt for solid samples presented in powdered form. This is common in near-IR spectroscopy. Particle size has a pronounced effect: decrease the particle size and the spectrum can be seen to shift bodily. This was seen to degrade the standard application of ridge regression in Fearn (1983), and was patched up by Hoerl et al. (1985) essentially by transforming the variables so that they add to zero, their row mean transformed model. If only contrast variables are used then one has the extreme version of the above prior with $\rho \to 1$.

The AR(1) covariance structure considered earlier in the chapter had two hyperparameters, as does the above intraclass covariance structure (k and ρ). One approach would be to specify a prior distribution for these parameters and then carry out a full Bayesian analysis, albeit with a need for numerical integration. Simpler approximate techniques are possible, eschewing this hyperprior specification, such as maximum integrated likelihood or cross-validatory choice as in Sections 6.8 and 6.9. This is also the preferred method of Fearn (1992).

A variant on the shrinkage to prespecified coordinate directions is provided by Oman (1991). He applies Stein shrinkage towards the space spanned by selected principal components of the design matrix.

A further and superficially rather attractive prior distribution for the near-IR data has been suggested by Fearn in the discussion of Brown and Mäkeläinen (1992). If Beer's law holds then the spectrum of the mixture should be the appropriate linear combination of the spectra of the pure components, that is,

$$Y = \sum_1^c z_r p_r = Zp$$

where $Y(q \times 1)$ is the spectrum of the mixture, p_r is the proportion of rth pure substance it contains, and $z_r(q \times 1)$ is the spectrum of the rth pure substance, $Z = (z_1, \ldots, z_c)$, $p = (p_1, \ldots, p_c)^T$. It cannot be assumed that one knows the proportions p, but if the mixtures are drawn from a population in which the covariance matrix of p is $V(c \times c)$, then the covariance of Y is the $q \times q$ matrix

$$M_{(q)(q)} = ZVZ^T.$$

Provided the spectra of the pure substances are available then the only unknown is the covariance matrix of the proportions, V. This, being ($c \times c$), is quite small ($c = 5$ for detergent data) compared with the induced covariance matrix of Y. Moreover, the spectrum of the mixture under this formulation is such that its mean spectrum will tend to peak and trough mimicking to some extent the spectrum of variances, a feature evident from plots 1.6 of sugar spectra and the overlaid detergent spectra in Fig. 7.8. Foreseeable limitations are the strict non-applicability of Beer's law. Beer's law holds only for substances in a medium, in small quantities and well dispersed to avoid interaction effects. The spectra of pure substances have to be calculated by subtraction of a medium. Also the medium for the mixture needs to itself be non-absorbant, for it is typically not present in small proportions. Water far from satisfies these conditions in the case of the aqueous solution of sugars as will be seen in Chapter 7.

   One advantage of the mixture prior formulation is that the dimension of the hyperparameter space is fixed as q tends to infinity. It may be seen that it consequently avoids the prior determinism elucidated by Dawid (1988) whereby one can almost surely predict the response with some finite subset of the regressors from the prior distribution alone.

6.11   Bayesian alternative to kriging

The body of this chapter has been concerned with prior distributions for regression when the set of explanatory variables may be viewed as a realization of a one-dimensional continuous random process. What if this is two-dimensional and thus constitutes the realization of a random field? Again, to promote robustness, the strong prior structure is imposed at the second (hyperparameter) level and not at the primary model level.

   The methodology offers an alternative to kriging in geostatistics, with the attractive feature that the linear predictor, the essence of kriging, has uncertainty incorporated into it. Here we present a somewhat incomplete but, it is hoped, instructive account. For further details refer to Le and Zidek (1992) and references therein. Multivariate extensions are given in Brown *et al.* (1993b).

   In a region we envisage a set of monitored or graded sites, g in number, together with a set of unmonitored or ungraded sites, u in number. There are assumed to be n observations at $p = u + g$ sites each observation having a set of k covariates, $z_j (k \times 1), j =, \ldots, n$, with values constant over sites within an observation. Actually observations at the u ungraded sites are not taken, but it is at least desired to predict the last (nth) observation at these ungraded sites.

   Specifically let $Y_j = (Y_j^{(1)T}, Y_j^{(2)T})$ be a p-dimensional random vector where $Y_j^{(1)}$ and $Y_j^{(2)}$ are u- and g-dimensional vectors, respectively ($p = u+g$). Assume that there are n independent vector observations $Y_1, \ldots, Y_n$ and one observation, the nth, is of future interest. Conditional on n values of the covariates the multivariate general linear model is assumed,

$$Y = ZB + E \qquad (6.15)$$

with $Y(n{\times}p)$ the random matrix of responses, $Z(n{\times}k)$ the matrix of known regressors, and the unknown matrix of regression coefficients $B(k \times p)$. The random error matrix $E(n \times p)$ is such that

$$E \sim \mathcal{N}(H, \Gamma) \qquad (6.16)$$

conditional on the $(p \times p)$ column covariance matrix Γ, subsequently regarded as unknown. Typically $H = I$ the $(n \times n)$ identity matrix. We assume the prior distribution on (B, Γ) to be the natural conjugate constructed as, given Γ,

$$B - B^0 \sim \mathcal{N}(H^0, \Gamma) \qquad (6.17)$$

and marginally

$$\Gamma \sim IW(\delta; \Psi), \quad \delta > 0, \qquad (6.18)$$

with hyperparameters the single shape parameter δ and the $p \times p$ scale matrix Ψ. Over the $p = u + g$ ungraded and graded sites this scale matrix is crucial to incorporating appropriate spatial information linking the ungraded to graded sites. The graded sites in turn allow estimation of parametrized unknowns in Ψ, perhaps resulting from some isometric structure as generated by the covariance kernel

$$\rho(\tau) = \exp -\alpha|\tau|^\kappa, \quad 0 < \kappa \le 2. \qquad (6.19)$$

This would reduce the $\frac{1}{2}p(p+1)$ parameters of Ψ to the two hyperparameters (α, κ) together with a third parameter denoting the common variance.

We adopt the rather simple approach to Bayesian analysis embodied in Section 6.2. This entails forming the prior distribution of the observable random variables, a matrix-variate T distribution, and then using standard conditional properties of the distribution as given in Appendix A. For an alternative derivation see Le and Zidek (1992). A first simplification is to note that given Γ

$$ZB \quad \sim \quad ZB^0 + \mathcal{N}(ZH^0Z^T, \Gamma)$$
$$E \quad \sim \quad \mathcal{N}(H, \Gamma)$$

and hence

$$Y \sim ZB^0 + \mathcal{N}(H + ZH^0Z^T, \Gamma). \qquad (6.20)$$

We have succeeded in eliminating the design term, ZB, in eqn (6.15). All that now needs to be done is to marginalize over the prior distribution of Γ and then use the conditional properties of the matrix-variate T to form needed posterior predictive distributions. This is more straightforward than the route advocated by Dawid (1981, p. 273).

The prior distribution of Y is such that with

$$Q = H + ZH^0Z^T$$

$$Y \sim ZB^0 + T(\delta; Q, \Psi). \tag{6.21}$$

Now splitting the response into ungraded and graded sites, $Y = (Y^{(1)}, Y^{(2)})$, let

$$\eta = \Psi_{22}^{-1}\Psi_{21}, \quad y^{(2)} = Y^{(2)} - ZB^{0(2)}.$$

The distribution of the $(n \times u)$ ungraded random matrix, given $Y^{(2)}$, is

$$Y^{(1)} - ZB^{0(1)} - y^{(2)}\eta \sim T(\delta + g;, Q + y^{(2)}\Psi_{22}^{-1}y^{(2)T}, \Psi_{11.2}). \tag{6.22}$$

The marginal distribution of the nth row of this random matrix $Y^{(1)}$ is, by virtue of Dawid's notation, just obtained from matching up elements of the matrix-T. The means of the vectors $Y_n^{(1)}(u \times 1)$ and $Y_n^{(2)}(g \times 1)$ are denoted

$$a_1 = z_n^T B^{0(1)}, \quad a_2 = z_n^T B^{0(2)},$$

respectively. The mean corrected vector of the graded sites is denoted

$$y_n^{(2)} = Y_n^{(2)} - a_2.$$

Marginally and given the data on the graded sites and $q_n = 1 + z_n^T H^0 z_n$,

$$Y_n^{(1)} - a_1 - \eta^T y_n^{(2)} \sim T(\delta + g; \Psi_{11.2}, q_n + y_n^{(2)T}\Psi_{22}^{-1}y_n^{(2)}). \tag{6.23}$$

This is a u-variate multivariate-T with $\delta + g$ degrees of freedom, again in the distributional notation of Appendix A. Note the special form of this predictive distribution for the ungraded sites: it does not depend on the graded data at other sites. There is thus no upgrading of the information. Such upgrading, the essence of the exercise of predicting ungraded sites, can only arise after a further level of prior information linking ungraded and graded sites together. This may be achieved by a simplified parametric form of Ψ as obtained say from eqn (6.19), the three unknown hyperparameters of this being estimated from the data over graded sites. The n observations on the graded sites may be examined marginally to estimate these hyperparameters separately by the methods alluded to earlier in Section 6.8.

7
Non-linearity and selection

7.1 Introduction

All the models so far considered have been linear in the unknown parameters. This allows the fitting of models non-linear in the measured variables, but at the expense of consequent non-linearity in prediction after controlled calibration. In this chapter we address these non-linear models, both from the viewpoint of exploring the extra complexities that arise and of reviewing methods for avoiding them. In particular variable selection techniques can be used to pick variables for which a linear model holds. This can be particularly advantageous in relating spectra to composition in that the spectral curve measured at a large number of wavelengths embodies a considerable degree of redundancy, and wavelengths may be selected at which the model relating absorbance to composition is quite linear. At other wavelengths which are avoided the relationship may be far from linear. We start in Section 7.2 by offering some diagnostics for detecting non-linearity and interaction and illustrate these on the sugars data described in Example 1.6. Then methods of variable selection are reviewed and illustrated on the same data as well as the detergent data of Example 1.5. Finally, properties of non-linear models relating to possible multimodality of inference are explored in Section 7.10. The earlier sections of the chapter focus on near-IR spectral data but the issues are quite general.

7.2 Plotting techniques for non-linearity and selectivity

In multicomponent infrared calibration, linearity is often assumed between composition and reflectance/absorbance. Beer's law may often be used as a justification. This assumed linearity may not hold at certain regions of the spectrum, however. Our first purpose is to develop methods to identify regions of the spectrum where linearity does not hold; secondly within the linear regions to indicate the components most favoured at each wavelength. These are achieved by simple graphical plots using the training data.

The data have composition fixed by design so that our preferred mode of analysis designates the reflectance spectra as the multivariate response (700 dimensions). The aim of the graphical techniques is to see whether there are bands of wavelengths where linearity according to Beer's law does not hold.

Beer's law states that reflectances or absorbances (logged absorptions) are linearly related to concentration, provided that the concentrations of ingredients are low, and that interactions between ingredients, and between ingredients and the solvent, are negligible. It applies to well dispersed ingredients in a non-absorbant medium. For the sugar data, the medium water is highly absorbant. Interactions between the ingredients and water (and possibly among the ingredients) may be considerable. Hence Beer's law may not hold.

The idea behind the plotting technique is simply to apply analysis of variance techniques separately to each of the 700 wavelengths in the spectrum, analysing the variance of absorbance across the 125 samples as explained by the three sugar factors each at five levels. If Beer's law holds then only the linear main effects will be significant. In reality a somewhat reduced version of this was implemented. Since the levels of the three sugars were 6, 10, 12, 14, and 18, we defined the linear effect contrast, -3, -1, 0, 1, 3 (x) and correspondingly the quadratic effect contrast, 5, -3, -4, -3, 5 ($x^2 - 4$). For interaction effects we confined our analysis to linear by linear terms. Thus nine degrees of freedom were used in fitting the model: a linear and a quadratic term for each sugar and three interaction terms; from the 124 available, after accounting for an overall mean.

Figure 7.1 gives the proportion of variation explained by these nine effects. Superimposed by vertical dotted lines are the positions of the 18 wavelengths selected by the method described in Section 7.5. Notice that there are regions, for example around 1900 nm, where only about 20 per cent of the variation is explained, and this is where water has a dominating absorbance. This is a region of the spectrum where non-linearity and interaction is present at a higher order than the quadratic and linear by linear interaction used in explanation. Also, from Fig. 1.6, this is where there is most variability in the spectrum, and might wrongly be the favoured region in some methods of analysis. As an aside we note that all the wavelengths selected in Section 7.8 are in positions of near 100 per cent explanation of variation.

The nine degrees of freedom include three quadratic and three linear by linear interaction effects in addition to the three linear main effects. Figure 7.2 breaks down the explained variation in Fig. 7.1 by (a) the linear effects (solid curve), (b) the linear plus quadratic effects (dotted curve). The residual amount from the upper limit of 1 gives the proportion of the variation explained by the linear by linear interaction. Clearly in the region of wavelength 1900 nm the variation explained has substantial quadratic and interaction effects as well as the higher order effects noted from Fig. 7.1. There are substantial quadratic effects just below 1500 nm also, where there is known to be another water absorbance band. It appears that, for the wavelengths selected in Section 7.8, most of the variation is explained by the linear main effects, and therefore that Beer's law holds.

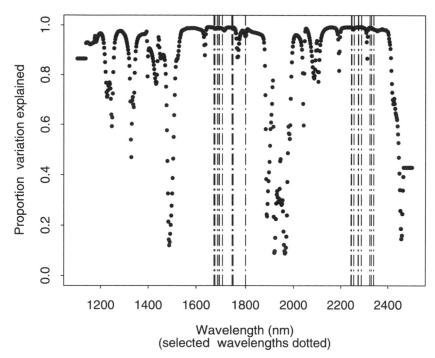

Fig. 7.1.   Sugar data, proportion of variation explained by linear, quadratic main effects and linear by linear interaction

A reduced version of Figs. 7.1 and 7.2 is given by Fig. 7.3 which just gives the variation explained by each of the three component linear effects. The lower broken curve is that variation explained by the linear effect of sucrose; the dotted curve above gives variation explained by linear effects of sucrose and glucose together; and the final solid curve above is the variation explained by all three linear effects together. The effect of fructose is then measured by the height of this solid curve above the dotted curve at each wavelength. The effect of glucose is similarly the height of the dotted curve above the broken curve at each wavelength. Because of the orthogonal design any other ordering of the successive fitting of sugars would produce the same effects for each sugar. This would not be the case if the design were not orthogonal. Then the measured effect in Fig. 7.3 of glucose would be *after fitting sucrose* and that of fructose *after fitting sucrose and glucose*.

Figures 7.4 and 7.5 are extracted from Figs. 7.1 and 7.2. They identify the largest of the three effects for all wavelengths where linearity is fairly assured. The two graphs have two different definitions of 'fairly assured'. The definition for Fig. 7.4 is less stringent than for Fig. 7.5. Figure 7.4 requires the explanation by the linear, quadratic and linear by linear in-

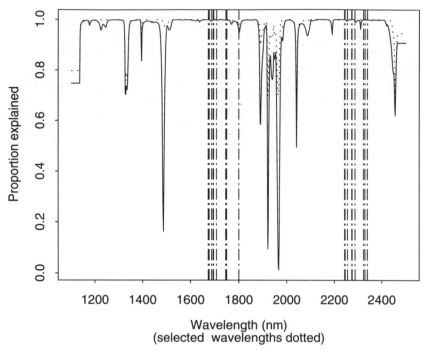

Fig. 7.2. Sugar data, split of variation explained: linear effects (solid); linear plus quadratic (dotted)

teraction of Fig. 7.1 to be at least 90 per cent, whereas in Fig. 7.5 it needs to be at least 98 per cent. Each figure also demands that there is at least 98 per cent attributable to linear effects in Fig. 7.2. The 'None' line of the figures gives those wavelengths where these assured linearity conditions do not hold. Figure 7.4 has 425 'linear' wavelengths (700 − 425 = 275 non-linear) whereas Fig. 7.5 comprises a subset of these, 189 in number, with 44 favouring sucrose, 46 glucose, and 99 fructose. Either of these two sets might represent a suitable screened set of wavelengths for further analysis. All but one (wavelength 1800 nm) of the 18 selected in Section 7.8 are included in the 189 wavelength set of Fig. 7.5.

The 425 wavelengths selected by the linearity screen displayed in Fig. 7.4 may contain much redundant information for prediction. The plotting and screening techniques above do not include covariance between reflectances at different wavelengths and so do not attempt to assess such communalities. Neither do the two plots, Figs. 7.4 and 7.5, attempt to identify the most selective wavelengths for each sugar component. They identify the sub-regions where they are likely to occur but it requires further processing such as envisaged later in this chapter to identify the most highly selective

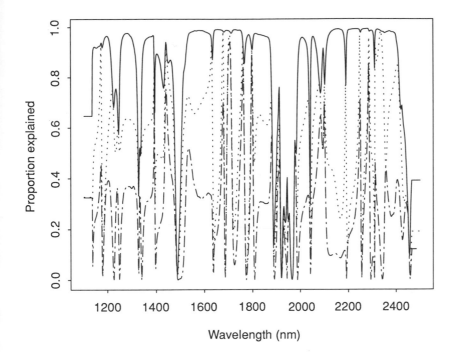

Fig. 7.3.   Variation explained by linear effects; sucrose; sucrose & glucose; sucrose & glucose & fructose

wavelengths. A mean-squared error of prediction of these linear screens are included with other selection procedures in Table 7.4 of Section 7.8.

Other methods of detecting non-linearity may be tailored to particular examples. One basic idea for a more formal procedure might be to hypothesize a model with non-linear terms in it. It would then be possible to test for a zero contribution from the non-linear part of the model. The use of just a few degrees of freedom for non-linearity and interaction is in the spirit of Tukey (1949). However, in the context of infrared calibration it is clear that you should avoid non-linear modelling by using only a carefully chosen subset of the wavelengths. In the next section variable selection methods are briefly reviewed and then a particularly simple method is introduced.

7.3   Variable selection with many variables

Variable selection can be beneficial for at least two reasons. Firstly a model, assumed linear in the explanatory variables, need not hold true for all the variables. Secondly even if the model is true, the paucity or structure of observations may mean that simple techniques like least-squares

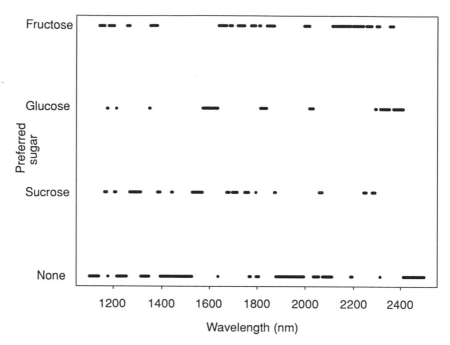

Fig. 7.4. Sugars with most of explained variance of absorbance at each wavelength. None denotes wavelengths of high non-linearity

and maximum likelihood produce unreliable estimates through sheer high dimensionality. The further difficulty with calibration problems is that if a model, fitted to training data, is linear in the sense of parameters and yet has non-linear explanatory terms such as x^2, then it will generate a model which is non-linear in the unknown composition parameter ξ in prediction. From a modelling viewpoint, including these terms is not without difficulties. The quadratic and linear terms need not provide a realistic model at other than the design points, and cannot be relied on for extrapolation or interpolation.

7.3.1   REGRESSION MODELS AND OVERFITTING

We wish to relate a set of variables, $Y(q \times 1)$ (the spectrum), to a set of variables $x(p \times 1)$ (the composition) when we have relatively few samples, n, of measurements on both sets of variables. It is not unusual for n to be very much less than q. In such indeterminate over-parametrized problems there are a variety of ways to regularize and achieve stable estimates: see Chapters 4–6. Whilst we may recognize that improvements are in principle possible with more information, since we are then always free

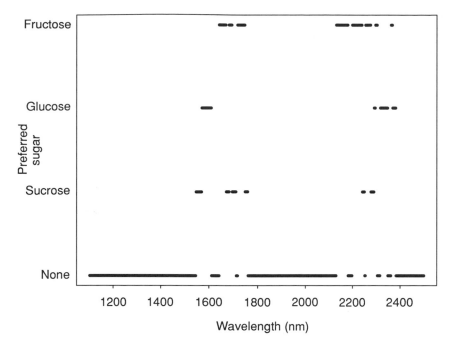

Fig. 7.5. Sugars with at least 98 per cent of explained variance of absorbance at each wavelength. None denotes wavelengths of high non-linearity

to ignore that extra information, there is a tradition, however, which emphasizes the fragility of inference with large numbers of variables and the vulnerability to overfitting. Even in the closed Bayesian inferential system, tractable prior distributions for large numbers of parameters have unforeseen and undesirable features, as discussed in Chapter 6. On the other hand, in the sampling framework of statistical inference, a set of complementary problems arise. A large number of estimates of effects which are truly zero will throw up at random a proportion of large estimates of seemingly large effects; the problem of multiple comparison. Regularization or shrinkage offers one answer, but may still be open to some overfitting. Internal leave-one-out cross-validation, a valuable tool against self-deception (see Chapter 4), may flatter in its internal cross-validated mean-squared error, especially as it assumes future prediction data that are very like that sampled. Later in an application of a technique of this chapter, using a validation set whose range is outside that of the calibrating data to predict the concentrations of three sugars in solution, the cross-validation mean-squared error is found to be an order of magnitude too small, chiefly here because of model inadequacy.

In one rather simple approach presented later in this chapter we will remove a large number of frequencies so as to reduce vulnerability to over-fitting. This is achieved by choosing frequencies to minimize the length of a confidence interval. Before specifying this method in Section 7.4 we review approaches to variable selection in the next section. The model adopted is presented in Section 7.4 and the application described in Section 7.6. The reader wishing to skip the technical detail of Section 7.4 may skip to the end of that section for a brief summary of the method before proceeding to the example.

7.3.2   VARIABLE SELECTION REVIEWED

The large number of responses and the relative paucity of observations in modern calibration make many of the standard variable selection techniques used in regression inapplicable or computationally prohibitive. For description of commonly used methods for multiple regression see Section 3.2.

The use of forward selection methods is widespread, and some instruments have such an approach incorporated in their accompanying software (see Osborne *et al.* (1984)). Fearn (1983) gives an example of calibrating protein content in ground wheat where forward selection is found wanting, partly because of high correlations between absorbances at different frequencies and low correlation with the component to be predicted. The use of stepwise regression also suffers in a similar way when applied to this example.

Selection techniques geared to prediction of the controlled variable via the fitted regression of Y on x have the same computational drawbacks. Brown (1982) utilizes a test of 'additional information' to select a subset of variables, and Spezzaferri (1985) utilizes a Bayesian information theoretic approach. Both these approaches are also applicable when $p > 1$.

Sometimes authors select principal components. This does not, however, select frequencies since all of the original set of frequency channels are still retained. Since model inadequacies appear at specific frequencies and not across the frequency range, selection of linear combinations of absorbances is unlikely to be entirely satisfactory.

In this chapter an extremely simple technique for response selection is described, based on Brown *et al.* (1991). It does not claim to give the smallest possible subset but it does provide a predictively good subset which substantially reduces the number of channels.

7.4   The model for selection

As discussed we adopt the approach of regression of the absorption spectra on composition. We also regress only on one ingredient at a time. Although in some respects it is desirable to regress on all five ingredients simultaneously, or at least on four of them since they total 100 per cent, for the initial stage of choice of frequencies we look to amalgamate the most discriminatory frequencies for each component taken in turn. We might add that ultimately we want to predict the concentration of each of the four detergent chemicals using just the spectrum without knowledge of concentrations of the other three chemicals in the sample and prediction equations for each component separately can be argued to be desirable (see Section 5.7.2). However, this is most likely to be successful if the equation can take account of the pollution effects of the other variables, and this is the reason why we seek an amalgamation of frequencies.

For the moment assume the simple linear regression models,

$$Y_{ij} = \mu_j + \beta_j x_k + \epsilon_{ij} \tag{7.1}$$

where $j = 1, \ldots, q$ indexes the $q = 1168$ frequencies for the detergent data, and $i = 1, \ldots, n$ indexes the $n = 12$ observations. Here x_i is the concentration of one of the ingredients, for the ith sample, centred to have mean zero. In this model, errors $\{\epsilon_{ij}\}$ have zero mean and variance σ_j^2 and are assumed to be uncorrelated. This is model (5.1) with one explanatory factor and a diagonal covariance matrix, $\Gamma = \text{Diag}(\sigma_1^2, \ldots, \sigma_q^2)$. The error variance at frequency j, σ_j^2 incorporates both measurement error and residual effects due to omitted constituents of the sample. The uncorrelatedness of the errors is certainly an over-simplification since correlation is induced by these omitted constituents. In this case it is possible to include the correlation of Y_{ij} across channels within an observation, but with the substantial additional problem of a very large unknown covariance matrix; see Denham and Brown (1993) for some possible ways of structuring such problems. A partial way around the problem is offered in Section 7.7. If such correlations are substantial it will be important to use them explicitly for prediction or to utilize them implicitly by a regularized method, such as partial least-squares, that regresses x on Y.

The n observations $\{Y_{ij}, x_i, j = 1, ..., q\}$, $i = 1, ..., n$ serve to estimate the unknown parameters, μ_j, β_j, σ_j^2 in model (7.1). With these q estimated calibrated relationships we may predict any further unknown sample compositions ξ_t, $t = 1, ..., T$, where we have designated unknown x by ξ (adjusted by the same amount as used to centre x). For this prediction problem the analogous model to eqn (7.1) is

$$Z_{tj} = \mu_j + \beta_j \xi_t + \epsilon_{tj} \tag{7.2}$$

where ϵ_{tj} are uncorrelated errors with variance σ_j^2, $j = 1, \ldots, q$, $t = 1, \ldots, T$.

One approach to choice of model dimension is to choose the size of model by calculating the mean-squared error of prediction. This mean-squared error is the sum of a variance and a bias squared term. Larger models have increased variance but decreased bias and an optimum compromise between the two is possible. A related but different approach which we will adopt focuses on prediction intervals. The fitted model from eqn (7.1) is not the true model. It is estimated once but may be used repeatedly to predict future values of ξ, each prediction incorporating the same bias of fitting. In addition to this source of error from the calibrating experiment there is the error generated by the post calibration or prediction experiment. These two sources of error are affected by the number of frequency channels (q) adopted and determine the width of the confidence interval. Although the methodology is formulated to deal with it, we do not here look at the simultaneous satisfaction of all future use prediction intervals with an ascribed probability; rather, we focus marginally on each single use interval. Again we are able to achieve an optimal compromise which minimizes the width of an interval and gives a unique choice of selected channels.

Quite distinct but problematic considerations arise within a Bayesian framework. Suppose $\xi_1, ..., \xi_t, ...$ are viewed as exchangeable *a priori*, that is having a joint distribution which is invariant to permutations of the indices, and consequently not in general independent; then $Z_1, ..., Z_t$ provide information on the probabilistic form of the exchangeability and posterior inference about ξ_t will depend on $Z_1, ..., Z_{t-1}$ as well as Z_t and the training data. Additionally, the multiple use of the calibration with its prior necessitates a careful assessment of the stability of inference to alterations of this prior distribution (see Berry (1988)).

For the selection of frequency channels in the next section we will work within the sampling theory inferential framework. Our basic idea is to choose that subset of q' of the q channels such that *the approximated length of each single use prediction interval is minimized*. Both q' and the corresponding subset of frequencies are chosen by the method.

7.5   The selection method

Since x and ξ are scalars we seek a linear combination of the response to each instrument. We can then use well developed univariate methodology. Henceforth in the prediction model eqn (7.2) we refer to the tth future Z and drop the subscript t to it. Let $Z = (Z_1, ..., Z_q)^T$ and $Z_\theta = \Sigma \theta_j Z_j$. Similarly let $\mu_\theta = \sum \theta_j \mu_j$ and $\beta_\theta = \sum \theta_j \beta_j$. We will look for values of $\theta = (\theta_1, ..., \theta_q)^T$, with typically $q' < q$ non-zero components, which minimize the approximate length of certain confidence intervals. In models (7.1), (7.2) in addition to the second-order error assumptions the errors are taken to be normally distributed. Throughout the development the error variances σ_j^2 are assumed known.

For prescribed θ, the compound response Z_θ is normally distributed with mean, the calibration line, $\theta^T E(Z) = \mu_\theta + \beta_\theta \xi$, and variance $\sum \theta_j^2 \sigma_j^2$. We proceed by means of the Cauchy–Schwarz inequality. Firstly, with prescribed probability $1 - \gamma$ and fixed θ with q' specific non-zero components,

$$\left| Z_\theta - \sum \theta_j(\mu_j + \beta_j\xi) \right| = \left| \sum \theta_j\sigma_j\epsilon_j^* \right| \le \sqrt{\sum \theta_j^2\sigma_j^2}\sqrt{\chi_{1-\gamma}^2(q')} \quad (7.3)$$

since the introduced ϵ_j^* are independent standard normal. Secondly and similarly, with probability $1 - \delta$

$$\left| \sum \theta_j(\hat{\mu}_j + \hat{\beta}_j\xi) - \sum \theta_j(\mu_j + \beta_j\xi) \right| \le \sqrt{\sum \theta_j^2\sigma_j^2 s^2(\xi)}\sqrt{\chi_{1-\delta}^2(q')} \quad (7.4)$$

where $s^2(\xi) = [1/n + \{(\xi - \bar{x})^2/\Sigma(x_k - \bar{x})^2\}]$ and q' is the number of prescribed non-zero θ_j. These statements will also be true for θ chosen by the calibrating data in model (7.1) provided the randomness induced does not effect the actual channels selected. We will give sufficient conditions for this later.

We let $c_1(q') = \sqrt{\chi_{1-\gamma}^2(q')}$ and $c_2(q') = \sqrt{\chi_{1-\delta}^2(q')}$. Now the triangle inequality gives $|Z_\theta - \hat{E}(Z_\theta)| \le |Z_\theta - E(Z_\theta)| + |E(Z_\theta) - \hat{E}(Z_\theta)|$, where $\hat{E}(Z_\theta) = \hat{\mu}_\theta + \hat{\beta}_\theta\xi$, and enables us to combine inequalities (7.3) and (7.4) into a single inequality for the divergence of Z_θ from its fitted value for given ξ. Solving this for ξ would give the single use confidence region. More simply, thinking of a graph of Z against ξ for a limited region of ξ values we may approximate the width of the confidence interval by 'height' divided by 'slope', essentially a local linear Taylor series expansion (see for example Carroll and Spiegelman (1986)).

For notational convenience we have assumed that the selected frequencies are the first q' out of q. The approximate half-width is thus

$$\left[c_1(q')\sqrt{\sum_1^{q'} \theta_j^2\sigma_j^2} + c_2(q')\sqrt{\sum_1^{q'} \theta_j^2\sigma_j^2 s^2(\xi)} \right] / \left| \sum_1^{q'} \theta_j\hat{\beta}_j \right| \quad (7.5)$$

This is the quantity we will minimize with respect to the q' channels with non-zero θ_j.

When σ_j^2 have to be estimated from the data they are replaced by the usual unbiased estimators $\hat{\sigma}_j^2$, but here no attempt has been made to adjust the probability statements accordingly.

The non-zero θ_j which minimize ratio (7.5) may be seen, by Lagrange's method, to be proportional to $\hat{\theta}_j$, where

$$\hat{\theta}_j = (\hat{\beta}_j/\sigma_j^2)/ \left(\sum_1^{q'} \hat{\beta}_j^2/\sigma_j^2 \right), \quad (7.6)$$

$j = 1, \ldots, q'$. Our estimator of ξ obtained from the linear compound of the

Z_j, $j = 1, \ldots, q$ obtained from eqn (7.6) is

$$\hat{\xi} = Z_{\hat{\theta}} - \hat{\mu}_{\hat{\theta}} \tag{7.7}$$

since $\hat{\beta}_{\hat{\theta}} = \sum \hat{\beta}_j \hat{\theta}_j = 1$.

Coincidentally, eqn (7.6) gives the generalized least-squares estimators conventionally used in multivariate controlled calibration, and eqn (7.7) is a special single component case of eqn (5.15) with diagonal covariance structure. The minimized half length of interval (7.5), substituted by eqn (7.6), is

$$[c_1(q') + c_2(q')s(\xi)] / \sqrt{\sum_1^{q'} \hat{\beta}_j^2/\sigma_j^2}. \tag{7.8}$$

It remains to select the number q'. With c_1 and c_2 functions of q' and linked by $s(\xi)$ in the numerator of half-width (7.8), our choice of q' and that subset of q' of the q channels, depends on the unknown ξ. However, if we choose equal probability levels $\gamma = \delta$ so that $c_1 = c_2 = c$ then minimization of half-width (7.8) over q' becomes independent of ξ. Moreover, whether or not the levels are chosen equal, half-width (7.8) is easily minimized by ordering the absolute values of the standardized slope coefficients $\hat{\beta}_j/\sigma_j$ (the signal-to-noise ratios) and choosing the largest q' of these, and then plotting half-width (7.8) by the number q'.

Sufficient conditions for there to be little variation in the channels selected are:

1. The variances σ_j^2 are small.
2. The slopes $\{\beta_j\}$ form two groups.

The first has absolute value far from zero, $|\beta_j| >> 0$. The second group has slopes approximately equal to zero. We need also that the slopes β_j in the first group are well spaced relative to the σ_j.

It is evident that the number of components selected is a monotonic decreasing function of the both γ and δ, so that if they are both chosen large enough only one component is selected. Typical values $\gamma = \delta = 0.1$ allow enough information to be retained in our detergent example. Throughout the sequel a common significance level $\gamma = \delta$ is chosen, and this common value is called η.

In summary, our method orders the absolute values of $\hat{\beta}_j/\sigma_j$, with $\hat{\sigma}_j$ estimating σ_j when the error standard deviation is unknown. It chooses the frequencies corresponding to the largest q' of these where q' minimizes half-width (7.8), and here $c_1^2(q')$ and $c_2^2(q')$ are tabulated chi-squared percentage points on q' degrees of freedom. Dependence of choice of the q' frequencies on $s(\xi)$ disappears if the two confidence levels are equal.

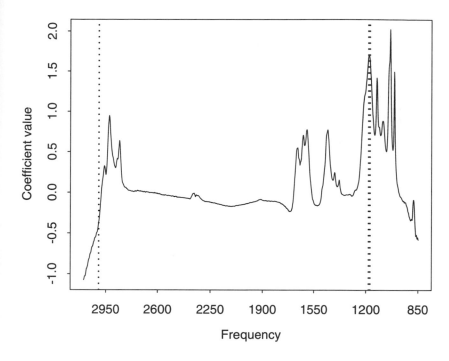

Fig. 7.6.   Detergent data, multiple regression of absorbance on compositions, component 1 coefficients by frequency, with component 1 selected frequency chains

7.6   The detergent example

The data consist of five components given in Table 1.4. The $n = 12$ spectra were presented in Figs. 1.5 and 6.1.

For component 1, Fig. 7.6 gives the bands of frequency channels selected for chosen common probability level $\eta = 0.1$. The positions of the chosen frequencies are indicated by vertical lines and are superimposed on the graph of slope coefficients for each component from multiple regression of absorbances on components. For this component there are eleven selected frequencies, in order {1176, 1174, 1178, 1180, 1173, 1182, 2997, 1171, 1184, 2999, 2995}, making up two quite separate groups of adjacent frequencies, {2999, 2997, 2995} and {1184-1171}. Figure 7.7 is proportional to the half-width (7.8) as a function of q' for component one, depicting the minimum at $q' = 11$ and a sharp increase as q' increases, with the value at the minimum being around one tenth of that with all 1168 frequencies included. Figure 7.7 is typical of plots of half-width (7.8) for the other components, and these have consequently been omitted, as have the other component plots paralleling Fig. 7.6. The number of frequencies selected

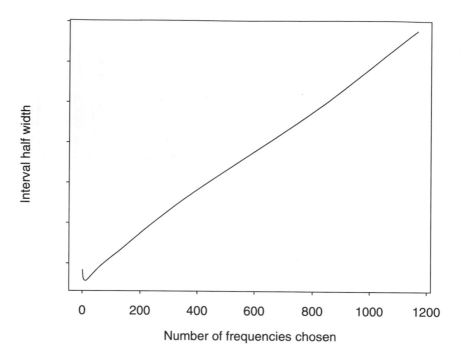

Fig. 7.7.   Detergent data, component 1 90 per cent confidence interval half-width by number of chosen frequencies

for components two to five are 9, 3, 7, and 17, in a further five bands. All the selected frequencies are quite distinct, so that the pooled set of frequencies for the hybrid multiparameter prediction involves $Q = \sum q' = 47$ selected frequencies. The positions of the seven bands are marked along the frequency axis of Fig. 7.8, and Fig. 7.9, which give overlays of the 12 spectra and the 12 with mean subtracted, respectively. This enables one to see where selections are relative to the mean curve and variation.

Table 7.1 gives leave-one-out root-mean-squared prediction errors for the five components corresponding to four different methods. The tabulated results are thus cross-validated so that the selection of wavelengths is based on eleven observations and the twelfth is predicted for every subset of size eleven. The first two methods incorporate all 1168 wavelengths and the second two involve $Q = 47$ wavelengths selected as indicated above. The two methods are further dichotomized by whether they are uni- or multi-component. The uni-component method uses estimator (7.7). The multi-component method uses four components to explain the variation in the Q absorbances; the fifth, water, is calculated by subtraction from 100 per cent. It then applies equation eqn (5.15) for prediction, with a diagonal

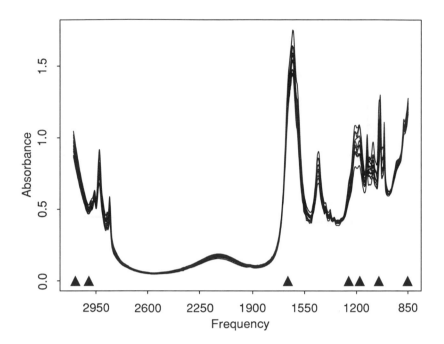

Fig. 7.8. Detergent data, twelve spectra overlaid with chosen frequency bands marked

Table 7.1.   Detergent data, mean-squared prediction error (cross-validated), 1168 or selected 47 frequencies

		Component			
Method	1	2	3	4	5
Simple LS	11.815	0.904	3.137	0.132	1.904
Multiple LS	0.154	0.217	0.159	0.013	0.091
Select Simple LS	2.336	0.184	0.655	0.155	0.878
Select Multiple LS	0.020	0.042	0.023	0.013	0.052

estimated covariance matrix $\hat{\Gamma}$ of error. As a consequence of the restriction to the diagonal covariance structure, there is a substantial reduction of prediction error in using multi-component methods and within either uni- or multi-component the selection procedures of this section are beneficial for most components. The variance of the four detergent ingredients and water

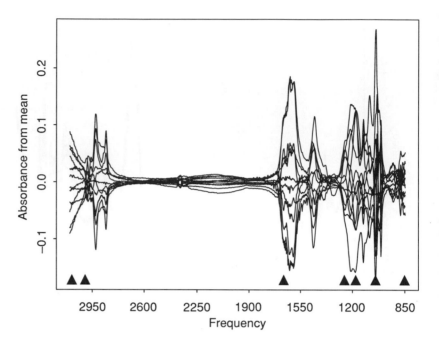

Fig. 7.9. Detergent data, twelve spectra differences from mean spectrum overlaid with frequency bands marked

were 15.0, 6.0, 3.6, 2.6, and 30.0, respectively so that predictions are generally very accurate with very high percentages of variation explained. As a further reference point, partial least-squares on all frequencies with four latent factors gave mean-squared prediction errors of 0.042, 0.083, 0.023, 0.011, and 0.033 for the four components and water, the second component results being given in Section 4.6.1. Whilst these partial least-squares results are only marginally less good than our preferred last row of Table 7.1 and may be improved by optimized cross-validatory choice of number of factors, 6, 5, 8, 7, and 9 respectively for the five components, the next section further extends the method and applies it to the calibration of sugars. It shows that partial least-squares on all frequencies lacks robustness with respect to different prediction sets. After selection of frequencies by developments of the method of this chapter, such non-robustness disappears.

The most obvious flaw with the methodology so far is the assumption of uncorrelated errors. This has two aspects: firstly the selection method needs to be adapted to account for these, and secondly the estimators as used in Table 7.1 ought to allow for the inclusion of correlation. The first of these is approached by transformation, whilst the second may be approached by the full use of generalized least-squares or other estimators discussed in Chapter 5.

7.7   Selection with autocorrelated errors

Often in the above approach, bands of adjacent wavelengths are chosen. We refer explicitly to the sugar data. The dominant correlation structure of the errors from model (7.1) within the $q = 700$ wavelengths is that of a stationary autoregressive process of order 1. That is the correlation between an error (ϵ_j) at wavelength j and one at wavelength $j - 1$ denoted by ρ, is constant as j varies over the 700 channels. For an AR(1) process these errors are generated as $\epsilon_j = \rho\epsilon_{j-1} + u_j$ where the $\{u_j\}$ are independent white noise. Consequently the covariance at lag t, $E(\epsilon_j\epsilon_{j-t})$, is proportional to ρ^t. Thus, if we transform the q-vector Y essentially by taking 'differences' $y_{ij} - \rho y_{i(j-1)}$, then the formulation will revert to the diagonal error structure of Section 7.5. The only modification required is that, when a 'wavelength' is chosen, in reality we are choosing both it and its immediately previous adjacent wavelength. (See Chapter 6 for further discussion of such transformations.) The AR(1) structure of a single spectrum is taken to be very approximate and of use in wavelength selection only: when it comes prediction from chosen wavelengths we prefer to be more cautious in our assumed covariance structure.

7.7.1   HYBRID SELECTION STRATEGY

We amalgamate the chosen wavelengths for the three sugars taken in turn. Sometimes hybrid strategies have been adopted:

(a) thinning out the chosen wavelengths when they are adjacent, taking the one with the largest absorption to standard error, in the uncorrelated case, $\rho = 0$;

(b) taking the first and last of a run of wavelengths when $\rho > 0$;

(c) taking the union of frequencies from $\rho = 0$ and the chosen $\rho = 0.9$.

Our broad strategy is in three steps:

1. Choice of wavelength set based on calibration data,
2. Construction of a predictor using calibration data,
3. Validation of the predictor on the prediction data.

In step 1 the usefulness of the subset is judged by leave-one-out (that is, cross-validated) mean-squared error using a uniquely specified simple predictor, for which we chose minimum length least-squares (see Chapter 4). Conceptually we might have preferred the generalized least-squares predictor from the regression of Y on a single explanatory x, using only q' wavelengths. When x has been centred and ξ adjusted conformably, then eqn (5.15) becomes

$$\hat{\xi} = \hat{\beta}^T\hat{\Gamma}^{-1}Z/(\hat{\beta}^T\hat{\Gamma}^{-1}\hat{\beta}) \tag{7.9}$$

where $\hat{\Gamma}$ is a $q' \times q'$ estimated unstructured covariance matrix formed from the least-squares residuals from model (7.1), restricted to the chosen q' wavelengths. This generalized least-squares (GLS) estimator is our preferred estimator for prediction in step 3, constructed in step 2 from the calibrating data, but with the amalgamated set of Q wavelengths. Step 1 allows you to compare different values of significance level η, differencing autocorrelation ρ, and hybrid strategies alluded to above.

Step 2 provides the predictor as estimated from the 125 observation data and step 3 validates the chosen predictors on the 21 observation validation data.

7.8 Application to sugars

Tables 7.2, 7.3, and 7.4 correspond to steps 1, 2, and 3, respectively, of Section 7.7.1. In Table 7.2 a selection of combinations of η and ρ are shown. Thus the first row of the table has confidence level $1 - \eta = 0.9$, no differencing ($\rho = 0$), and uses the 125 observation training data to first predict sucrose. The algorithm selects wavelengths 1708, 1706, and 1710 nm (in order). To predict glucose, wavelengths 2340 and 2342 nm are selected and, for fructose, channels 1688 and 2256 nm. Thus there are a total of 7 wavelengths or channels for the prediction of the three sugars. The corresponding leave-one-out mean-squared errors using multiple regression of the three sugars in turn on all seven wavelengths are 0.123, 0.236, and 0.201. Moving on to the next row of the table, increasing the confidence level to 0.99 adds further wavelengths to those already selected in the first row, giving an amalgamated total set of 12 wavelengths. The second block in the table incorporates differencing with $\rho = 0.9$. Because of differencing the wavelengths selected are thus pairs of adjacent wavelengths, doubling the number selected unless channels selected form a sequence. For the $\eta = 0.001$ row the amalgamated set of wavelengths is $\{1674, 1676, 1678, 1694, 1696, 1746, 1748, 1750, 2244, 2246, 2274, 2276, 2324, 2326, 2328, 2330\}$, a total of sixteen wavelengths for regression prediction. These, when thinned as in Table 7.3, are reduced to 12 (removing 1676, 1748, 2326, 2328). The preferred value of η is 0.001 and ρ of 0.9. Various other combinations were investigated although only a very selective presentation is given in Table 7.2. A hybrid strategy involving the union of wavelengths manually chosen from $\rho = 0$ and $\rho = 0.9$ with thinning is given in the last block of Table 7.2, as a selection of 18 wavelengths for multiple regression prediction, as are also used in the last row of Table 7.3.

Table 7.3 gives the leave-one-out cross-validated mean-squared errors at the optimum numbers of factors for partial least-squares regression of concentration on absorptions. This is also given for the full 700 wavelengths spectra.

Table 7.4 gives the mean-squared error for the predictors on the 21

Table 7.2. Calibration data, probability level η, autocorrelation ρ, total number of channels selected, Q, and mean-squared error (MSE) for three sugars, with leave-one-out predictions by minimum length least-squares multiple regression (MLLS)

			MSE (MLLS)		
η	ρ	Q	Sucrose	Glucose	Fructose
0.1	0	7	0.123	0.236	0.201
0.01	0	12	0.118	0.236	0.152
0.001	0	15	0.102	0.227	0.124
0.001	0.9	16	0.079	0.150	0.112
0.0001	0.9	23	0.083	0.150	0.112
0.001	0.5	18	0.087	0.166	0.131
0.001	0 & 0.9	18	0.075	0.148	0.105

observation validation set, all models estimated on the 125 observation calibration data. The partial least-squares regression predictors correspond to a best number of factors for leave-one-out predictors applied to the calibration data. Minimum length least-squares (MLLS) is applied to all 700 wavelengths. Both the generalized least-squares predictor given by eqn (7.9) and partial least-squares are applied to the 18 wavelengths selected here. Also included are two selections by the linear screen of Section 7.2, namely the 425 wavelengths with at most 10 per cent non-linearity, the 189 with at most 2 per cent non-linearity. The final column gives the average of the mean-squared error for the three sugars.

There is a clear and dramatic message. Based on an internal leave-one-out cross-validation criterion internal to the calibration data, you would be inclined to think that selection was not beneficial. However, when we move from Tables 7.2 and 7.3 to Table 7.4 we see that there is a dramatic deterioration of the full 700 wavelength predictors whilst the selected wavelength predictors show only a modest deterioration. Selection is clearly beneficial for prediction of all three sugars, markedly so for sucrose and fructose. The linearity screens also give an improvement over using the full 700 wavelengths, with greatest improvement for the most stringent screen of 2 per cent non-linearity. We attribute the beneficial effect of selection as due to its removal of those wavelengths where Beer's law has broken down and there are substantial non-linearities.

7.9   Commentary

In the above no claim is made to have chosen the unique best subset of variables for prediction. The claim is rather that the strategy chooses

Table 7.3. Calibration data, optimum PLS predictors and MLLS and leave-one-out mean-squared error for given sets of wavelengths

Wavelengths	ρ	Method	Mean-squared error		
			Sucrose	Glucose	Fructose
700	–	MLLS	0.259	0.292	0.503
700	–	PLS	0.078	0.140	0.144
12	0.9	PLS	0.082	0.149	0.106
18	0 & 0.9	PLS	0.085	0.152	0.107

Table 7.4. 21 observation validation data, mean-squared error for predicting sugar components, models estimated on calibration data

Wavelengths	Method	Mean-squared error			
		Sucrose	Glucose	Fructose	All
700	MLLS	211.1	33.8	748.3	331.1
700	PLS	2.72	0.31	3.95	2.33
425	PLS	0.33	1.96	0.60	0.96
189	PLS	0.35	0.85	0.28	0.49
12	PLS	0.23	0.75	0.27	0.41
18	PLS	0.12	0.41	0.26	0.26
18	GLS	0.23	0.22	0.36	0.27

good small subsets which are predictively better than the complete set of wavelengths. The theory of Section 7.5 deals only with uncorrelated error structures. The transformation through a fractional differencing of Section 7.7 is an attempt to get around this, as are the hybrid strategies for thinning out the selected wavelengths.

The positions of the selected wavelengths given as vertical chains in Figs. 7.1, and 7.2 may be compared with the mean and standard deviation curve for the sugars data given in Fig. 1.6. It may be noted, perhaps surprisingly at least for adherents of principal component regression, that the chosen wavelengths correspond neither to high mean nor to high variance of absorbance. On further analysis this is not surprising and embodies the strength of the method. This exposes the corresponding weakness of PCR even when principal component choice is based on correlation with composition and not pure variance. The original directions are determined by maxima of the variance. The dominant and spoiling role of the spectrum of water with its non-linear effects has been bypassed by selection. This is because of their inflationary effect on σ_j as estimated from model (7.1) for wavelengths j with sizeable non-linearity or interaction effects. The presence of these effects was demonstrated in Section 7.2.

The methods above cope with $p > 1$ components by considering each component versus the rest of the components. You may, however, envisage situations where the selection of frequencies should be based on a completely multi-component method. This would be needed if, for example, particular frequencies were very good at discriminating between either of two components against the rest, but offered little discrimination between the pair, whose resolution could be achieved from other channels.

It is easy to see qualitatively how you might consider components simultaneously, and at the same time incorporate a correlation structure across frequencies. The ordering of wavelengths above may be justified from an information standpoint. The asymptotic (as $n \to \infty$) covariance matrix of $\hat{\xi}$, the generalized least-squares estimator, is given in Chapter 5 as the $p \times p$ matrix $(B\Gamma^{-1}B^T)^{-1}$ (see Section 5.5). Here B is a $p \times q$ matrix of coefficients from the full version of model (7.1) formed by regressing the q absorptions on the p components. The information matrix of the maximum likelihood estimator, with the same asymptotics, is just the inverse of this covariance matrix. It is of interest to note that when $p = 1$ and the $q \times q$ covariance matrix Γ is diagonal then the criterion just amounts to accumulating the q' highest information components, with B and Γ estimated by least-squares. Thus the ordering based on $\hat{\beta}_j^2/(\hat{\sigma}_j^2)^2$ is quite natural. The use of information in itself does not give you a stopping rule which adequately takes account of the dimensionality of estimation. However, the stopping criterion of Section 7.5 has some assumptions built into it and in the sugar example of this chapter *a priori* choice has been supplanted by cross-validatory choice of confidence level.

Figure 1.6 indicates that there are small but important differences between the calibration and prediction spectral data. The prediction set is almost a full replicate of a balanced design with the same middle factor level of 12, so that the mean spectra are fairly alike. When using a spectrum from a prediction set, unusualness of that spectrum may be checked in ways important to prediction following Section 5.11.

7.10   Non-linear theory

Methods for detecting non-linearity of a regression relationship were explored in Section 7.2. Methods for selecting variables so as to avoid non-linearity have subsequently been discussed in Sections 7.3–7.7. We now go back to fundamentals and explore the difficulties that may arise in non-linear modelling, where now we are concerned about non-linearity in unknown parameters. We assume throughout the following model for a

q-variate response vector $Z = (Z_1, ... Z_q)^T$, where

$$Z = h(\xi) + \epsilon, \ E(\epsilon) = 0, \ \text{Cov}(\epsilon) = \Gamma \tag{7.10}$$

and $h(\xi) = (h_1(\xi), ..., h_q(\xi))^T$ is a non-linear function of the unknown scalar ξ. It will suffice to only consider scalar ξ to encapsulate many of the difficulties of non-linear h. The error vector ϵ is taken to be multivariate normal. For this model to represent standard univariate regression we would require Γ diagonal and q would then denote the number of uncorrelated observations. Our perspective is, however, that of multivariate calibration in which ξ represents the true unknown value of the covariate of interest, and typically the function $h(.)$ is estimated from training data in which observations Z_i are taken at controlled prespecified x_i, $i = 1, ..., n$. Our aim then with $h(.)$ estimated by $\hat{h}(.)$ is to predict an unknown ξ from an observed Z as given by eqn (7.10). Assume that possible values of ξ lie in an interval I. We may represent the expectation of Z, that is $h(\xi)$, as a curve in q-dimensional space, as the planar ($q = 2$) quadratic curve in Fig. 7.10. As ξ varies it traces out a curve in q dimensions, the locus of the mean of Z. Any point Z may also be represented as a point in the q-dimensional figure. Thus in Fig. 7.10 the axes may be taken to be Z_1 and Z_2, the mean locus is the curve given, and two possible archtypal Z points 1, 2 are marked; the dotted region is the set of double points as developed in the sequel. The asymmetry of the figure is a consequence of the asymmetry of the interval I of ξ-values. The behaviour of inference about ξ is also quite different for points 1 and 2 for a different reason. Had the locus been a line then no difference of behaviour would be seen. In this section these differences are explored. The curvature of the mean locus in Fig. 7.10 is independent of parametrization of ξ, that is it does not disappear by taking ξ^2 or some function of ξ. Thus the work of this section follows intrinsic curvature measures, as opposed to parameter effects curvature. The distinction is discussed by Beale (1960) and Bates and Watts (1988).

For simplicity assume that the training data provide perfect information on h and Γ so that we may in (1) transform Γ to the identity matrix. Had the training data been finite, and $h(.)$ of the form given as

$$h_i(\xi) = \beta_{i0} + \sum_{k=1}^{K} \beta_{ik} g_k(\xi), \ i = 1, ..., q,$$

then $h(.)$ and Γ could be replaced by generalized least-squares estimates $\hat{h}(.)$ and $\hat{\Gamma}$. There may then be a need to adjust inferential procedures for the finiteness of the sample.

With Γ the identity matrix, to estimate ξ we minimize

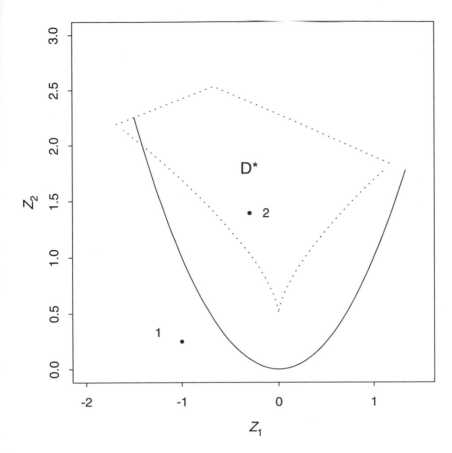

Fig. 7.10. Parabolic locus, $Z_2 = Z_1^2$ with double region for $-2 < Z_1 < 1.5$

$$f_Z(\xi) = ||Z - h(\xi)||^2, \ \xi \in I.$$

That is, we choose $\hat{\xi}$ to be that value of ξ such that the distance from Z to $h(\xi)$ is a minimum. Under normal errors this will be maximum likelihood when h is known; otherwise it has a generalized least-squares justification. Other distance functions would be implied by other non-normal error assumptions.

The squared distance function, $f_Z(\xi)$, can provide both a point estimator of ξ and confidence sets, for example the ξ-set such that $f_Z(\xi) < c$. Several problems may arise:

(i) The function $f_Z(\xi)$ may have more than one local minimum.

(ii) Our confidence set may end up being several disjoint intervals.

The definition of a problematic point Z adopted in Brown and Oman (1991) is that a *double point* is any Z such that f_Z has two (or more) local minima.

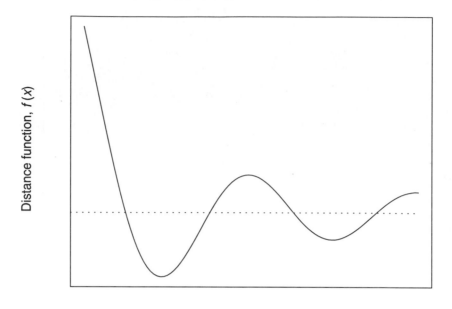

Fig. 7.11.   Distance function with two minima

Thus Z is double if it gives rise to the plot of f_Z versus ξ in Fig. 7.11. Formally they define the set of double points $D^*(\xi_1, \xi_2)$:

$$D^*(\xi_1, \xi_2) = \{Z \epsilon R^q : f'_Z(\xi_i) = 0,\ f''_Z(\xi_i) > 0,\ i = 1, 2,\ \xi_i \in I,\ \xi_1 \neq \xi_2\}$$
$$(7.11)$$

and

$$D^* = \cup D^*(\xi_1, \xi_2).$$

Brown and Oman (1991) use $q = 2$ variables. Using differential geometry, they characterize double points given by eqn (7.11) in Z-space. The double points Y are such that

(a) they lie at the intersection of at least two normals at ξ_1 and ξ_2 on $h(\xi)$, and

(b) the signed Euclidean distance from $h(\xi_i)$ to Z, measured in the direction of the normal, is at most the radius of curvature at $h(\xi_i), i = 1, 2$. Additionally in the case of likelihood confidence sets:

(c) the modulus of the difference of the squared distances be at most γ, the chi-squared upper 100α per cent critical point on one degree of freedom; and that one of ξ_1 and ξ_2 is the global minimum.

Here (a), (b), and (c) define a $(1 - \alpha)$-double point.

In designating regions of double points the *evolute* has a crucial role since it is the locus of centres of curvature, and tangents to the evolute curve are normals to $h(\xi)$. The set of double points D^* for the quadratic locus of Fig. 7.10 and an asymmetric interval of possible ξ values is superimposed in the same figure as the region inside the closed dotted boundary. Brown and Oman (1990) extend the discussion to higher dimensions.

Perhaps a few remarks are in order about why double points are important. The existence of more than one local minimum of $f_Z(\xi)$ does not of course rule out a global minimum, but is problematic for at least two reasons. Firstly, it may pose problems for any numerical minimization routine, and several starting values may be necessary. Secondly, inferentially it suggests that a single point-estimate may be at best misleading. This inferential aspect is reinforced when you turn to confidence sets where you may be uncomfortable in communicating disconnected intervals for ξ, even though they reflect the ambiguities of support for different values of ξ.

For a given ξ, a double point will be more or less likely to occur depending on its Euclidean distance from $h(\xi)$. As a summary measure of the extent of the set of double points we may calculate $p(\xi)$, where

$$p(\xi) = P(Z \in D^*),$$

is the probability of Z being a double point for a given value of ξ.

Such a measure of the extent of the double points for a particular ξ is intended to offer diagnostic insight into the relationship between Z and ξ. For a Bayesian it can at most be for design or preposterior purposes. The Bayesian may after all note that if he or she observes a particular Z, the likelihood and also the posterior distribution can be plotted and summarized and any multimodality reported. In the design context, however, he or she may wish to give average rates for pathologies over future Z. This can be of interest in itself or for choosing between competing Z variables when they are available.

8
Pattern recognition

8.1 Introduction

This chapter explores the use of Bayes theorem in simple applications of pattern recognition and identification. The main development that follows is of discrimination between p multivariate normal populations. This introductory section concludes with two examples on updating evidence in applications of medical screening and of forensic science.

In the training data the p populations are identified by p dummy variables, 1 for the presence of that population, 0 for absence. For each observation in the training data these constitute the x-variables corresponding to q-variate Y. Once a linear relationship between x and Y is established, in future it is desired to predict the x (population of origin) from an observed q-vector Z (the future Y). This is classical discrimination, and directly parallels the calibration paradigm of the rest of this book. In particular, recall the direct link between the regression of Y on x and that of x on Y of Section 5.5. This algebraic link does not depend on whether x is a random variable. In particular there is the well-known two-population result that linear discrimination as developed by Fisher can be reproduced by regression of the dummy variable, which identifies the population of origin, on the q variates. However, there is one special feature here: the x-variables are identifiers of the population and not continuous variates. It is therefore desirable and possible to further give the probability that the future Z derives from each of the p populations.

Our development makes no attempt to review all the work in multivariate discrimination. In one respect it does, however, go further than is usually allowed in multivariate textbooks. By introducing proper prior distributions, it is possible to discriminate when there are more variables than observations. Such discrimination, as often arises using spectroscopic data, is particularly important in, for example, the pharmaceutical industry, where it is essential to confirm rapidly that the correct ingredients are used at each stage of production.

Bayes theorem will be used in two ways. Firstly the simple discrete version for giving the $p(C_j \mid D)$, $j = 1, \ldots, p$ from $p(C_i)$ and $p(D \mid C_i)$, $i = 1, \ldots, p$, where C_i denotes the event that the future observation comes from population class i. The $\{C_i\}$ are exclusive and exhaustive, and D denotes the data, the q-vector Z. As a direct consequence of the definition of conditional probability, Bayes theorem gives

$$p(C_j \mid D) = p(D \mid C_j)p(C_j)/\left[\sum_{i=1}^{p}(D \mid C_i)p(C_i)\right]. \qquad (8.1)$$

An alternative form of this is $p(C_j \mid D) \propto p(D \mid C_j)p(C_j)$ or

$$p(C_j \mid D)/p(C_i \mid D) = [p(D \mid C_j)/p(D \mid C_i)][p(C_j)/p(C_i)], \qquad (8.2)$$

and if there are just two populations, so that $C_j = C, C_i = \bar{C}$, then eqn (8.2) reads: posterior odds equals likelihood ratio times prior odds.

The second way in which Bayes theorem is used is in forming $p(D \mid C_i)$. Typically the distribution of Z involves unknown parameters (population means and covariance matrices). The posterior distribution of these parameters is available from the training data, and $p(D \mid C_i)$ may be formed as a posterior predictive average.

Before proceeding to this development of discrimination, the power and implications of Bayes theorem are illustrated in two simple examples, one from medical screening, the other from forensic science.

1. This example concerning the mandatory testing for HIV, the viral precursor to AIDS, derives from Gastwirth (1987). The figures used here are probably not inappropriate more recently, but distinctly more accurate genetic tests now exist, so the example is not meant to reflect the situation now.

   Although several moral issues are raised, we shall concentrate on the statistical problems. An estimate of the number of US citizens having the HIV virus is 1.5 million:– let us say 6 per 1000. Suppose that for an individual with the virus, the probability of a *positive* outcome to the test is 0.998 (the sensitivity of the test); suppose that for an individual without the virus the probability of a *negative* outcome is 0.98 (the specificity of the test). What is the probability that someone with a positive test has the virus?

   Applying Bayes theorem as in eqn (8.2),

   $$p(HIV \mid +)/p(H\bar{I}V \mid +) = (0.998/0.02)(0.006/0.994) = 0.301,$$

   so that $p(HIV \mid +) = 0.231$, and there is only a 23 per cent chance that the diagnosed individual is truly HIV positive. Fallacious non-probabilistic arguments focus on 0.998 or the smaller of (0.998,0.98) as the probability that an individual is truly HIV. The low probability of HIV coming from Bayes theorem derives from the 2 per cent of a rather large pool of non-HIV individuals.

2. A murder has been committed in an isolated community. Identification evidence links a suspect with the scene of the crime. What is the probability that he is guilty?

Forms of evidence might be: a match between fingerprints found on the murder weapon and those of the accused; a correspondence between DNA found in blood-stains at the scene, and that of the accused; rifling marks on a bullet, which could have been produced by the accused's gun; characteristics of an incriminating typed letter, which match those of letters typed on a certain type-writer; or eye-witness evidence of age, sex, height, and so on, which describe the accused.

Here the accused individual is either guilty or innocent. Let D be the event that the accused possesses a characteristic (the identification evidence) which otherwise occurs in a general member of the population with (typically small) probability P. Let π be the probability of guilt taking all other evidence into account, and assume that this evidence and the identification evidence are conditionally independent, given either innocence or guilt. Now $p(D \mid \text{guilty}) = 1$, $p(D \mid \text{innocent}) = P$; and the likelihood ratio in favour of guilt is $1/P$, the prior odds are $\pi/(1 - \pi)$, and applying Bayes theorem yields,

$$p(\text{guilty} \mid \text{all evidence}) = \pi/(\pi + P - P\pi).$$

Suppose $P = 0.01, \pi = 0.001$ then this probability of guilt is still only 9.1 per cent. On the other hand, a naive non-probabilistic argument, 'the prosecutor's fallacy', identifies $p(\text{innocent} \mid D)$ with $p(D \mid \text{innocent})$ yielding $p(\text{guilty} \mid D) = 1 - P$.

This formulation is taken from Dawid (1993a); see his paper for further development and see also the references therein. An earlier discussion of the celebrated case, The People versus Collins, is given by Cornfield (1969).

8.2 Discrimination with unequal covariances

Our development assumes multivariate normal populations with differing unknown q-vector means. In this section the covariance matrices are assumed unknown and different from population to population. Section 8.3 restricts to a common unknown covariance matrix.

Suppose the jth population has mean vector $\mu_j (q \times 1)$ and covariance matrix $\Sigma_j (q \times q)$. We assume that the p sets of parameters (μ_j, Σ_j), $j = 1, \ldots, p$ are a priori independent. Thus with training data consisting of n_j observations independently from each population, $j = 1, \ldots, p$, we may treat the p populations separately. For the discussion here a Normal–Inverse–Wishart prior is assumed for (μ_j, Σ_j). Our training data is described by the model

$$Y_j - 1_j \mu_j^T \sim \mathcal{N}(I, \Sigma_j),$$

with $Y_j (n_j \times q)$ and $1_j (n_j \times 1)$ is a vector of ones. The prior distribution is such that, given Σ_j,

$$\mu_j^T - m_j^T \sim \mathcal{N}(h_j, \Sigma_j),$$

and

$$\Sigma_j \sim \mathcal{IW}(\delta_j; Q_j).$$

Then *a posteriori* from Appendix B.1.1, given Σ_j,

$$\mu_j^T - \tilde{\mu}_j^T \sim \mathcal{N}([1/h_j + n_j]^{-1}, \Sigma_j) \tag{8.3}$$

where $\tilde{\mu}_j = w_j m_j + (1 - w_j)\bar{Y}_j$, $w_j = 1/(1 + h_j n_j)$.

Suppose a q-vector Z comes from this jth population, then given (μ_j, Σ_j),

$$Z^T - \mu_j^T \sim \mathcal{N}(1, \Sigma_j)$$

and the right-hand side does not involve μ_j so that this holds conditional only on Σ_j, and the equation may be added to eqn (8.3), giving

$$Z^T - \tilde{\mu}_j^T \sim \mathcal{N}(1 + (1/h_j + n_j)^{-1}, \Sigma_j). \tag{8.4}$$

Furthermore the posterior distribution of Σ_j from the training data is $\mathcal{IW}(\delta_j^*; Q_j^*)$ where $\delta_j^* = \delta_j + n_j$,

$$Q_j^* = Q_j + S_j + (\bar{Y}_j - m_j)^T [h_j + 1/n_j]^{-1}(\bar{Y}_j - m_j),$$

and $S_j = (Y - 1_j \bar{Y}_j)^T (Y - 1_j \bar{Y}_j)$. Using this posterior inverse Wishart to remove the conditioning in eqn (8.4) and the characterization of the matrix-variate Student T (strictly multivariate-T here) from Appendix A,

$$Z^T - \tilde{\mu}_j^T \sim T(\delta_j^*, a_j, Q_j^*)$$

with $a_j = 1 + (1/h_j + n_j)^{-1}$.

Finally suppose before observing Z the probability that the observation comes from population j is π_j, then Bayes theorem, eqn (8.2), yields

$$p(C_j \mid Z) \propto \pi_j g_j^* a_j^{\delta_j^*/2} |Q_j^*|^{-1/2} \{a_j + (Z - \tilde{\mu}_j)(Q_j^*)^{-1}(Z - \tilde{\mu}_j)\}^{-(\delta_j^* + q)/2}, \tag{8.5}$$

with $g_j^* = \Gamma\{(\delta_j^* + q)/2\}/\Gamma(\delta_j^*/2)$. This allows calculation of $p(C_j \mid Z)$, where C_j is the event that the observation comes from population j. In addition to the training data, the prior ingredients are $(m_j, h_j, \delta_j, Q_j)$, $j =$

$1, \ldots, p$. Various levels of elaboration and simplification are possible. A prior for these prior parameters could be assigned, linking together the parameters of the different populations. Moving in the direction of simplification rather than elaboration, it might not be unreasonable to assume vague prior knowledge about the means. This is achieved by letting $h_j \to \infty$, $j = 1, \ldots, p$. If we are concerned about applications with $q > n_j$ then it is desirable to retain a proper prior distribution for Σ_j by retaining $\delta > 0$. Further simplifications are then possible to the assumed structure of the $(q \times q)$ matrix Q_j. The smallest number of hyperparameters would arise from $Q_j = kI$, reminiscent of ridge regression smoothing (see Friedman (1989) and Campbell (1980) for similar approaches to regularized discrimination, but without our Bayesian roots). It would be a degree less restrictive to assume an intraclass covariance structure, favoured by Lindley *et al.* (1985). If $\delta_j = \delta$ common to all populations, then the ridge form would require the specification of two hyperparameters, δ, k and the intraclass form, three hyperparameters. These could be simply estimated as plug-in modal values (maximum integrated likelihood estimates) or be estimated by cross-validation as discussed in Chapter 6. Other more general structures for Q are suggested by Chen (1979) and estimated by the EM (expectation-maximization) algorithm of Dempster *et al.* (1977), regarding the prior distribution as incomplete information about Σ_j.

The distinction between controlled and random calibration manifests itself in the specification of the prior probability, π_j, that the future observation comes from population j. No information on π_j is furnished by controlled calibration. Previous experience alone must be used to specify π_j, and such experience may suggest an estimate that bears no relationship to $n_j / \sum n_i$. On the other hand, when the future Z comes from the jth population according to the same random mechanism that generated n_j training observations, then the n_j can serve to update any previous experience; perhaps such experience is in the form of a Dirichlet prior distribution for $(\pi_1, \ldots, \pi_p)$. For a large number of training observations $n_j / \sum n_i$ would not be an unreasonable estimate for π_j, although worries would arise if the training data were thought to be generated by a different selection mechanism from the future observations.

Multivariate Normal Bayesian discrimination was first developed by Geisser (1964) using vague prior distributions. Dawid and Fang (1992) have explored the determinism inherent in the natural conjugate approach as $q \to \infty$. Brown *et al.* (1993a) suggest estimating a covariance matrix using a (partially conjugate) generalized inverse Wishart prior distribution. Brown (1976, 1980) has explored other difficulties with the Dirichlet natural conjugate to the multinomial used in simple Bayesian medical diagnosis. The points are further developed by Dawid (1993b) in the context of selection bias.

In the next section we explore the standard restriction that the p covariance matrices are equal.

8.3   Discrimination with a common covariance

It is assumed that the p covariance matrices of the p multivariate normal populations have the same covariance matrix Σ. Intervening cases are possible between this and completely different matrices assumed in the previous section, and have been alluded to, for example by assuming the Σ_j form a random sample. We will not explore this further, but see Smith and Spiegelhalter (1981) for a compromise between common and different covariance matrices, but in the context of vague prior distributions.

Again the Bayesian natural conjugate approach amounts to a straightforward application of Appendix B.1.1. This time because of the common covariance matrix all p training sets are combined as $Y(n \times q)$ and $X(n \times p)$ where $n = \sum_{i=1}^{p} n_i$ and X consists of p dummy vectors identifying the population of origin of the observation. Thus $X^T X = \mathrm{Diag}(n_1, \ldots, n_p)$. To conform exactly with the notation of Appendix B.1.1 the unknown matrix $B(p \times q)$ denotes the p different population mean q-vectors. Assigning the same Normal–Inverse–Wishart prior as in the Appendix, the posterior distribution is as given and it just remains to develop the predictive distribution of a q-vector Z emanating from the jth population. Let $e_j(p \times 1)$ be the indicator vector, 1 in the jth position, 0 otherwise. Then given B and Σ,

$$Z^T - e_j^T B \sim \mathcal{N}(1, \Sigma). \tag{8.6}$$

The right-hand side does not involve B, so that the statement is true conditional only on Σ. Also from the appendix, the posterior distribution of B given Σ is such that

$$e_j^T (B - \tilde{B}) \sim \mathcal{N}(e_j^T (H_0^{-1} + X^T X)^{-1} e_j, \Sigma),$$

with $\tilde{B}$ the posterior mean of B. Adding this to eqn (8.6) gives, conditional on Σ,

$$Z^T - e_j^T \tilde{B} \sim \mathcal{N}(a_j, \Sigma) \tag{8.7}$$

where $a_j = 1 + e_j^T (H_0^{-1} + X^T X)^{-1} e_j$. Also *a posteriori* $\Sigma \sim \mathcal{IW}(\delta^*; Q^*)$ with $\delta^* = \delta + n$ and $Q^* = Q + S + (\hat{B} - B_0)^T (H_0 + (X^T X)^{-1})^{-1} (\hat{B} - B_0)$, with $S = (Y - X\hat{B})^T (Y - X\hat{B})$, the pooled sum of products from the sample means of each population. Hence removing the conditioning on Σ eqn (8.7) becomes

$$Z^T - e_j^T \tilde{B} \sim \mathcal{T}(\delta^*; a_j, Q^*),$$

and from the form of the multivariate Student density given in Appendix A the posterior probability that Z emanates from the jth population is given by

$$P(C_j \mid Z) \propto \pi_j a_j^{\delta^*/2} \{a_j + (Z^T - e_j^T \tilde{B})(Q^*)^{-1}(Z - \tilde{B}^T e_j)\}^{-(\delta^*+q)/2}. \quad (8.8)$$

That is, Z is judged by its distance from $e_j^T \tilde{B}$, the posterior mean of the jth population in the metric defined by Q^*. Posterior probability eqn (8.8) also varies with π_j and a_j (determined by n_j, the number of training observations in the jth population). The prior parameters are δ, Q, B_0, and H_0. If $H_0 \to \infty$ so that the information about B is vague then $a_j = 1 + 1/n_j$, $Q^* = Q + S$, $\tilde{B} = \hat{B}$, the matrix of sample means, and only prior parameters δ, Q need be specified. As in the previous section these may be further elaborated or restricted.

Some insight into the form of the region for which $p(C_j \mid Z)$ is highest may be obtained by letting $n_j \to \infty$, $j = 1, \ldots, p$. Then in the limit we have the known parameter multivariate normal rule,

$$p(C_j \mid Z) \propto \pi_j \exp[(-1/2)(Z - \mu_j)^T \Sigma^{-1}(Z - \mu_j)$$

where $B^T = (\mu_1, \ldots, \mu_p)$ and we have used $Q^*/n \to \Sigma$ in probability, and $(1 + x/n)^n \to e^x$. Hence assign to population j rather than i if

$$\begin{aligned} log(\pi_j/\pi_i) \quad > \quad & (1/2)[(Z - \mu_j)^T \Sigma^{-1}(Z - \mu_j) - (Z - \mu_i)^T \Sigma^{-1}(Z - \mu_i)] \\ = \quad & [Z - (\mu_i + \mu_j)/2]^T \Sigma^{-1}(\mu_i - \mu_j). \quad (8.9) \end{aligned}$$

This is linear in the q-vector Z, and the regions are bounded by hyperplanes in q-dimensions. The regions are determined by the means μ_i, $i = 1, \ldots, p$ and the covariance matrix Σ. Using the symmetric square root of Σ and transforming to $Z^0 = \Sigma^{-1/2} Z, \mu_i^0 = \Sigma^{-1/2} \mu_i$, then the rule becomes ascribe to j rather than i if

$$log(\pi_j/\pi_i) > (1/2)[\|Z^0 - \mu_j^0\|^2 - \|Z^0 - \mu_i^0\|^2]$$

and the rule depends on the Euclidean distance of Z^0 to each of the two transformed means. The dividing hyperplane is the perpendicular to the line joining (and extending from) the two means, division depending on $log(\pi_j/\pi_i)$ and bisecting when $\pi_j = \pi_i$. For p populations and $q > p$, one need only consider the $(p-1)$ dimensional hyperplane that defines the p means (less than $(p-1)$ if linearly dependent). The p dividing hyperplanes meet in a point. For example with $p = 3$, $\pi_1 = \pi_2 = \pi_3$, the q dimensions may be reduced to a plane and the perpendicular bisectors of the sides of the triangle of means meet in a point. Transforming back the original coordinate system, the boundaries remain hyperplanes that meet in a point, but orthogonality is defined by $a^T \Sigma b = 0$, when a and b are conjugate axes of the ellipsoid determined by Σ. Conjugate axes both go though the origin, but one is parallel to the tangent to the other. This geometry of boundaries may also show that some regions are far from all the population means, and an observation in such a region should not really be allocated; rather it does not fit any population.

The linear discriminator (8.9), with least-squares estimates of μ_j and Σ is otherwise known as Fisher's linear discriminant function for two populations (see Fisher (1936)). It was motivated as that linear combination of the q variables for which the between to within group variance is maximized.

A rather different insight comes from rewriting eqn (8.8) for two populations as

$$p(C \mid Z) = 1/(1 + ae^{Z^T b}), \quad p(\bar{C} \mid Z) = 1 - p(C \mid Z),$$

so that the probability has a logistic form. The same form arises for multivariate (exponential family) distributions of Z other than the normal. Rather than assume multivariate normality of Z one can work with a binomial likelihood conditional on Z. This is more robust but shows some loss of efficiency when Z is multivariate normal (see Efron (1975)).

Obviously good discrimination is only possible when population means are well separated relative to the error covariance matrix Σ. (It is easy to calculate the probability of misclassification when all parameters are considered known.)

We end with a salutary warning concerning the danger of incorrectly relating the separation of the means to the error covariance structure. This is taken from Imperial College MSc 1972, and was there presented precisely to test the student's understanding of such dangers.

In a book on multivariate analysis, an example is given of principal component analysis for two bivariate normal samples A and B, say. Each sample is of size twenty-four, and the sample average vectors are

$$\bar{y}_A = (22.86, 24.40), \quad \bar{y}_B = (21.82, 22.84).$$

The two sample covariance matrices are found to be very similar and principal components are calculated from the pooled unbiased covariance estimate $\hat{\Sigma} = S/(n-1)$. The eigenvalues of $\hat{\Sigma}$ are $\lambda_1 = 40.80$ and $\lambda_2 = 0.50$ with corresponding eigenvectors, $v_1^T = (0.66, 0.75)$ and $v_2 = (-0.75, 0.66)$. The author states that 'we have indicated that the second of the two (principal components) is of little importance, so we can summarize the positions (of the two populations) on the first principal axis.' Using this single axis it is established that the two populations do not differ, and hence that the two populations are indistinguishable using y.

[The point is that small eigenvalues of the error covariance matrix offer the potential for good discrimination if the means are different in the corresponding eigenvector directions.]

8.4   Bibliography

Many multivariate textbooks provide standard material for normal theory discrimination or pattern recognition (sometimes also called supervised learning). References that are a little more special are Gnanadesikan (1977, Chapter 4), with his emphasis on graphical techniques, and Dempster (1969) for his geometric insights. An interesting recent approach, multivariate regression with p dummy variables as responses, using the relationship between the two regressions as in Section 5.5, and flexible estimation, with attendant S software, is presented by Hastie *et al.* (1993).

Appendix A
Distribution theory

A.1 Multivariate normal

Let $Y(q \times 1)$ be a random vector with density function $f(y)$. Then Y is said to follow a non-singular q-variate normal distribution with mean θ and positive definite covariance matrix $\Gamma(q \times q)$ if

$$f(y) = c(q)|\Gamma|^{-1/2} \exp\{(-1/2)(y - \theta)^T \Gamma^{-1}(y - \theta)\} \qquad \text{(A.1)}$$

with $c(q) = (2\pi)^{-q/2}$. If Γ is singular the Y is said to have a *singular* or *degenerate* distribution and the density does not exist except in the sense of lying on a hyperplane in q-dimensional space; probability calculations may still be carried out in this subspace. Often the shorthand notation

$$Y \sim N_q(\theta, \Gamma)$$

is used to denote that Y is distributed as a q-variate normal random vector with mean θ and covariance matrix,

$$\text{Cov}(Y) = E(Y - \theta)(Y - \theta)^T = \Gamma.$$

Linear transformations of normal vectors are again normal with

$$\text{Cov}(AY) = A\text{Cov}(Y)A^T$$

when $A(p \times q)$ is a matrix of constants. Sub-vectors of Y are also normal so that if $Y^T = (Y_1^T, Y_2^T)$ and conformably $\theta^T = (\theta_1^T, \theta_2^T)$,

$$\Gamma = \begin{pmatrix} \Gamma_{11} & \Gamma_{12} \\ \Gamma_{21} & \Gamma_{22} \end{pmatrix}$$

then

$$Y_1 \sim N_p(\theta_1, \Gamma_{11}).$$

The conditional distribution of Y_1 given $Y_2 = y_2$ is

$$Y_1|Y_2 = y_2 \ \sim \ N_p(\theta_1 + \Gamma_{12}\Gamma_{22}^{-1}(y_2 - \theta_2), \Gamma_{11.2}) \qquad \text{(A.2)}$$

where $\Gamma_{11.2} = \Gamma_{11} - \Gamma_{12}\Gamma_{22}^{-1}\Gamma_{21}$.

A.2 Matrix normal

Let $Y(p \times q)$ be a random matrix where every element has mean zero. Then the random matrix is said to have a matrix normal distribution, $\mathcal{N}(P, Q)$ if $p_{ii}Q$ and $q_{jj}P$ are the covariance matrices of the ith row and the jth column, respectively. This notation follows that of Dawid (1981); his non-standard notation is distinguished by the use of calligraphic capitals in distributional shorthand. The use of zero mean is not restrictive for if Y has mean matrix M then

$$Y - M \sim \mathcal{N}(P, Q).$$

Properties:

1. If $X \sim \mathcal{N}(R, Q)$ and independently $Y \sim \mathcal{N}(P, Q)$ then $X + Y \sim \mathcal{N}(R + P, Q)$.
2. Similarly if $X \sim \mathcal{N}(P, R)$ and independently $Y \sim \mathcal{N}(P, Q)$ then $X + Y \sim \mathcal{N}(P, R + Q)$.

These results follow from the matrix results

$$
\begin{aligned}
R \otimes Q + P \otimes Q &= (P + R) \otimes Q \\
P \otimes R + P \otimes Q &= P \otimes (R + Q)
\end{aligned}
$$

but note that $R \otimes S + P \otimes Q$ is not of a Kronecker product form.

Provided P and Q are positive definite, the matrix normal may be represented by the probability density function

$$f_Y(y) = c(p, q)|P|^{-q/2}|Q|^{-p/2}\exp\{(-1/2)\text{trace}[P^{-1}YQ^{-1}Y^T]\},$$

with $c(p, q) = (2\pi)^{-pq/2}$.

A.3 Wishart distribution

Let

$$X \sim \mathcal{N}(I_\nu, Q)$$

for Q non-negative definite, and let $V(q \times q)$ be a symmetric non-negative definite random matrix defined by

$$V = X^T X.$$

Then V is said to have a Wishart distribution with degrees of freedom, ν, and scale matrix $Q(q \times q)$, and denoted

$$V \sim W(\nu; Q).$$

When Q is positive definite and $\nu > q$ then V is positive definite almost surely and the density function exists and is given by

$$f(V) = c(q,\nu)|V|^{(\nu-q-1)/2}|Q|^{-\nu/2}\exp\{-(1/2)\text{trace}[Q^{-1}V]\}, V > 0, Q > 0,$$
(A.3)

with $c(q,\nu) = 2^{-q\nu/2}/\Gamma_q(\nu/2)$ and

$$\Gamma_q(t) = \pi^{q(q-1)/4}\prod_{i=1}^{q}\Gamma[t - (i-1)/2].$$

The Wishart distribution is a multivariate generalization of the chi-squared as is evident from the characterization above. Also by properties of the characterizing normal, square principal sub-matrices of V are also Wishart with the same degrees of freedom and scale matrix the conforming sub-matrix of Q. Moments are also readily derived from the characterization.

A.4   Inverse Wishart distribution

Here again the notation follows Dawid (1981) and diverges from the standard notation, the difference lying in the specification of degrees of freedom. Let $V(q \times q) \sim W(\nu; Q^*)$ with $Q^* > 0$, $\nu \geq q$, so that V is positive definite almost surely; then the distribution of $U = V^{-1}$ is Inverse Wishart with shape parameter $\delta = \nu - q + 1$ and is written

$$U \sim \mathcal{IW}(\delta; Q),$$

where $Q = (Q^*)^{-1}$. For $\delta > 0$ the density function is defined as

$$f(U) = c(q,\delta)|Q|^{(\delta+q-1)/2}|U|^{-(\delta+2q)/2}\exp\{(-1/2)\text{trace}(U^{-1}Q)\}, \quad U > 0,$$
(A.4)

with $c(q,\delta) = 2^{-q(\delta+q-1)/2}/\Gamma_q[(\delta+q-1)/2]$. In the more standard notation (see for example Press (1982)) the 'degrees of freedom' would be $m = \nu + q + 1 = \delta + 2q$, whereas Anderson (1984, p. 268) uses the superficially more natural ν of the Wishart distribution. We have resisted calling δ the degrees of freedom; that term more naturally remains with $\delta + q - 1$. With our definition of shape parameter for $\delta > 2$, we have that

$$E(U) = Q/(\delta - 2).$$

The distribution is of particular interest in normal theory regression analysis. The special form defined is especially easy to manipulate without having to constantly adjust the shape parameter. The Inverse Wishart distribution as defined is also consistent under marginalization, a property not shared by the other notations. A further property of extendibility is utilized in Chapter 6. It is also possible to define singular versions of the Inverse Wishart distribution.

A.5   Matrix Student T-distribution

Again the notation follows Dawid (1981), diverging from the standard notation in its assignment of degrees of freedom and scale. Let Φ be $\mathcal{IW}(\delta; Q)$ and, given Φ, $T \sim \mathcal{N}(P, \Phi)$; then the induced marginal distribution for $T(p \times q)$ is a matrix-T distribution denoted $\mathcal{T}(\delta; P, Q)$. It is clear also that $T^T \sim \mathcal{T}(\delta; Q, P)$. If $P = I_p$, $Q = I_q$ then the distribution $\mathcal{T}(\delta; I_p, I_q)$ is called the standard matrix-T distribution. If T is standard matrix-T then $P^{1/2} T Q^{1/2}$ is $\mathcal{T}(\delta; P, Q)$. For $p = q = 1$, $\mathcal{T}(\delta, 1, 1) = \delta^{-1/2} t_\delta$, where t_δ is the univariate Student t-distribution on δ degrees of freedom: to eliminate this discrepancy, an additional scale factor $\delta^{1/2}$ could have been carried along in the matrix definition, but would complicate formulae. The notation differs from that of Dickey (1967) and Press (1982) also in the matrix arguments P and Q, both of which for us rather naturally refer to covariance matrices, firstly for rows and secondly for columns.

The density function of the matrix-T exists if $\delta > 0$, $P > 0$, $Q > 0$ and for $T \sim \mathcal{T}(\delta; P, Q)$ is given by

$$f(T) = c(p, q, \delta)|P|^{(\delta+p-1)/2}|Q|^{-p/2}|P + TQ^{-1}T^T|^{-(\delta+p+q-1)/2}, \quad (A.5)$$

with

$$c(p, q, \delta) = \pi^{-pq/2} \Gamma_q[(\delta + p + q - 1)/2] / \Gamma_q[(\delta + q - 1)/2].$$

Marginal and conditional distributions of the matrix-T are also matrix-T. For example, if T is partitioned into $T^T = (T_1^T, T_2^T)$ where $T_i(p_i \times q)$, $i = 1, 2$, then marginally

$$T_2 \sim \mathcal{T}(\delta; P_{22}, Q).$$

Marginalization does not affect degrees of freedom as defined here. The form of the vector- or multivariate-T density is defined by putting $p_2 = 1$ when the power of the quadratic in $T(1 \times q)$ in (A.5) becomes $\delta + q$. Again the definition of the multivariate T corresponds to the standard definition except for a $\delta^{-1/2}$ scale factor. In this multivariate case the constant of proportionality in density (A.5) simplifies to

$$c(1, q, \delta) = \pi^{-q/2} \Gamma[(\delta + q)/2] / \Gamma(\delta/2).$$

Here the compacted notation $\mathcal{T}(\delta; Q)$ will often be used in place of $\mathcal{T}(\delta; 1, Q)$. The conditional distribution of T_1 given $T_2 = t_2$ is such that

$$(T_1 - P_{12} P_{22}^{-1} t_2) \sim \mathcal{T}(\delta + p_2; P_{11.2}, Q + t_2^T P_{11}^{-1} t_2).$$

Although derived under the assumption of multivariate normality, the multivariate Student distribution, constructed as it is through ratios or angles, depends on much weaker spherical symmetry and not normality (Dawid, 1977).

A.6 Matrix Fisher F-distribution

Suppose $U \sim W(\nu; \Phi)$ given Φ, where marginally $\Phi \sim \mathcal{IW}(\delta; I_q)$ with $\delta > 0$ and $\nu > q - 1$ or ν integral. The induced marginal distribution of U is called (by Dawid) matrix-F and denoted $U \sim \mathcal{F}(\nu, \delta; I_q)$. For $q = 1$ the distribution becomes $(\nu/\delta)F_\delta^\nu$ so that again some scale parameters are discarded from the traditional definition. From the construction of the matrix-T above, if $T \sim T(\delta; I_p, I_q)$ then $T^T T \sim W(p; \Phi)$ given Φ, with $\Phi \sim \mathcal{IW}(\delta; I_q)$. That is, $T^T T \sim \mathcal{F}(p, \delta; I_q)$.

Appendix B
Conditional inference

B.1   Some Bayesian results

In this appendix some results for Bayesian prior to posterior analysis are given in the context of the multivariate and matrix-variate normal. The notation of Appendix A is used.

B.1.1   POSTERIOR DISTRIBUTION OF A REGRESSION MATRIX

Suppose $Y(n \times q)$ and $X(n \times p)$ are observed as realizations of the model with given B,

$$Y - XB \sim \mathcal{N}(I, \Sigma),$$

with $\Sigma(q \times q)$ a temporarily assumed known covariance matrix. *A priori* assume given Σ that

$$B - B_0 \sim \mathcal{N}(H_0, \Sigma).$$

Then, given Σ, Y, and X, *a posteriori*

$$B - [WB_0 + (I - W)\hat{B}] \sim \mathcal{N}((H_0^{-1} + X^T X)^{-1}, \Sigma) \qquad \text{(B.1)}$$

where

$$W = (H_0^{-1} + X^T X)^{-1} H_0^{-1}.$$

Thus the posterior mean of B is a matrix weighted average of the prior mean B_0 and the least-squares estimate

$$\hat{B} = (X^T X)^{-1} X^T Y. \qquad \text{(B.2)}$$

Note that modification of the original multivariate regression model to incorporate a covariance across observations is included in the above result since

$$Y - XB \sim \mathcal{N}(H, \Sigma)$$

implies

$$H^{-1/2} Y - (H^{-1/2} X)B \sim \mathcal{N}(I, \Sigma),$$

and in the posterior distribution for B,

$$X^T X \to X^T H^{-1} X, \quad X^T Y \to X^T H^{-1} Y.$$

If in addition Σ is unknown and *a priori*

$$\Sigma \sim \mathcal{IW}(\delta; Q)$$

then *a posteriori*,

$$\Sigma \sim \mathcal{IW}(\delta + n;\; S + Q + (\hat{B} - B_0)^T \{ H_0 + (X^T X)^{-1} \}^{-1} (\hat{B} - B_0) \quad \text{(B.3)}$$

where

$$S = (Y - X\hat{B})^T (Y - X\hat{B}),$$

the least-squares residual error sum of products matrix. These results follow directly, after some algebra, from Bayes theorem and the form of the probability density functions.

B.2   Ancillarity and Bayesian analysis

The sampling notion of ancillarity has been developed from the work of Fisher to ensure that the sampling distributions used for classical inference are appropriately conditioned. There are well-known deficiencies with inferences which always use the unconditional model of observation generation; see for example Cox and Hinkley (1974, p. 35). By conditioning the classical statistician is unable to go as far as the Bayesian who conditions on all the data observed, for then there would be no sampling variation to utilize. Usually she will condition on *ancillary* statistics. These are, however, hard to set up unambiguously, since a given model may have several suitable ancillary statistics (see Basu (1964)). Putting aside such difficulties suppose $U = (T, C)$ is the minimal sufficient statistic for the model with parameters $\theta = (\psi, \lambda)$, being interest and nuisance vector parameters, respectively, with product parameter space $\Omega_\theta = \Omega_\psi \times \Omega_\lambda$. Then if the sampling density of U is such that

$$f(U|\theta) = f_1(T|C, \psi) f_2(C|\lambda) \qquad \text{(B.4)}$$

then C is said to be S-ancillary (S for Sverdrup and Sandved) for the nuisance parameter λ. Note in passing that requirement (B.4) is weaker than requiring that the distribution of the ancillary should not involve any of the parameters.

   If the sampling model has such a structure there are implications for the structure of a Bayesian analysis. Suppose the prior distribution for θ is such that ψ and λ are *a priori* independent, that is

$$\pi(\theta) = \pi_\psi(\psi) \pi_\lambda(\lambda).$$

The marginal posterior distribution of the interest parameter vector ψ from a standard Bayesian analysis, writing down the likelihood and multiplying by the prior distribution, is the same as would obtain had the likelihood been

$$f_1(T|C, \psi).$$

The Bayesian need not specialize to the minimal sufficient statistic and derive its distribution though; all that is required is the appropriate factorization of the full likelihood function. Such a factorization occurs with models from the exponential family in a mixed parametrization involving expected value parameter (λ) and canonical parameters (ψ) if and only if marginal distributions of C form an exponential family; see Barndorff-Nielsen (1978, p. 178) for a more formal statement. For example the independent Poisson model for a contingency table admits two sets of parameters, the marginal probabilities and the log odds ratios, the former are moment parameters and the latter canonical parameters, and C are also Poisson. A particularly important example is provided by the multivariate normal model. In regression it justifies treating the explanatory variables as if they were fixed even though in truth they are realizations of random variables. Suppose

$$X \sim \mathcal{N}(I, \Sigma)$$

with the random matrix $X(n \times p) = (X_1, X_2)$, a partitioning by column with $p = p_1 + p_2$ and X_1, X_2 $n \times p_1$, $n \times p_2$, respectively. Then

$$\begin{aligned} X_1|X_2 &\sim X_2 B^T + \mathcal{N}(I, \Sigma_{11.2}) \\ X_2 &\sim \mathcal{N}(I, \Sigma_{22}) \end{aligned}$$

with $\lambda = \Sigma_{22}, \psi = (B, \Sigma_{11.2})$. Moreover with *a priori* $\Sigma \sim \mathcal{IW}(\delta; Q)$, then the required *a priori* independence property of λ and ψ is assured, this being a consequence of conjugacy of the prior distribution: it is as if it were itself constructed from previous normal data. We list further points to note:

1. In terms of prediction one may obtain the following result. Suppose X is partitioned by column as above. Suppose $Y(n \times q)$ and $X(n \times p)$ are jointly normal given Σ $[(q + p) \times (q + p)]$. Additionally, assume $\Sigma \sim \mathcal{IW}(\delta; M)$. To predict Y^* given X_1^*, X, Y, suppose given Σ

   $$(Y^*, X^*) \sim \mathcal{N}(I_t, \Sigma)$$

   a further random sample of t observations on $(q + p)$ variables, conditionally independent of (Y, X) given Σ. Then unconditionally

   $$Y^*|X_1^*, X, Y \sim Y^*|X_1^*, X_1, Y.$$

Thus with only X_1^* for prediction of Y^* then the information provided by X_2 in the training data may be discarded.

2. Other notions of ancillarity are sometimes used (see Barndorff-Nielsen (1978)) although S-ancillarity seems the most natural extended notion, at least for Bayesians. An insightful discussion is given by Dawid (1975).

Appendix C
Regularization dominance

This appendix establishes sufficient conditions for when a member of a particular class of regularized estimators dominates least-squares. In doing this an unbiased estimator of the risk gain is obtained. The approach is a simplified univariate multiple regression version of results derived by Brown and Zidek (1980), which in turn adapts and extends Thisted (1976). The basic method of proof uses an integration by parts technique of Stein (1973). The derivations assume a full rank linear model, but could be adapted to the singular case provided the loss function were zero in the null space. The class includes a variety of shrinkage estimators: the original James–Stein estimator; this estimator adapted to multiple regression; and simple adaptive ridge estimators, like those given by the ridge parameter estimators $\hat{k}_{MHKB}$ and $\hat{k}_{MLW}$ of Section 4.3. In the context of ridge regression it specifies what can be achieved when the estimation of the ridge constant is incorporated in the overall procedure. Copas (1983) emphasizes the importance of shrinkage estimators to control for selection effects.

The class of estimators is any regularized estimator of the form

$$\hat{\alpha}_{iR} = (1 - \hat{w}_i)\hat{\alpha}_i, \qquad (C.1)$$

where $\hat{\alpha}_i = \hat{\alpha}_{iLS}$, the least-squares estimator of the regression p-vector of coefficients from the canonical model (4.8, 4.9), with $\alpha = V^T\beta$. Here the estimated weight is of the form

$$\hat{w}_i = t_i h_i \omega_i / \left(c_i + \sum_1^p \omega_j \hat{\alpha}_j^2 \right), \qquad (C.2)$$

with $h_i = \lambda_i^{-1}$, the inverse of the ith eigenvalue of $X^T X$. The non-negative constants t_i, ω_i, and c_i specify the chosen regularizing estimator. The form of (C.2) is that of the inverse of a quadratic form in the least-squares estimator.

The estimator is judged by expected quadratic loss, weighted across components,

$$\mathcal{R}(\hat{\alpha}_R) = E\left\{\sum_1^p L_i(\hat{\alpha}_{iR} - \alpha_i)^2\right\} \tag{C.3}$$

This incorporates the two cases of estimation and prediction loss as $L_i = 1$ and $L_i = \lambda_i$, respectively, as discussed after eqn (4.5).

Assume throughout that $p \geq 3$, since otherwise the integration by parts argument used below breaks down, and the result is in fact not true.

From (C.1) and (C.2) and the risk definition (C.3),

$$\mathcal{R}(\hat{\alpha}_R) - \mathcal{R}(\hat{\alpha}) = \sum L_i h_i E\{U_i^2 \hat{w}_i^2 - 2U_i \hat{w}_i(U_i - \lambda_i^{1/2}\alpha_i)\}. \tag{C.4}$$

Now, since the density function of the normal distribution of $(U_i - \lambda_i^{1/2}\alpha_i)$ is the exponential of the square in this term, the trick of Stein (1973) is to absorb this linear multiplier term into the probability measure by means of a simple integration by parts. Thus (C.4) becomes

$$E\left[\sum L_i h_i \{U_i^2 \hat{w}_i^2 - \partial(U_i \hat{w}_i)/\partial U_i\}\right].$$

With the form of (C.2) in mind let

$$\hat{w}_i = \{C_i(U)\}^{-1}$$

from which this risk difference becomes

$$E\sum L_i h_i \{(U_i/C_i)^2 - 2/C_i + 2(U_i/C_i^2)\partial C_i/\partial U_i\}.$$

Partially differentiating with the explicit form of the inverse of (C.2) for $C_i(U)$ gives the risk difference as the expectation of

$$\sum L_i h_i \{(U_i/C_i)^2(1 + 4/t_i) - 2/C_i\}. \tag{C.5}$$

This is then an unbiased estimator of the increase in risk of the regularized estimator over least-squares.

For the regularized estimator $\hat{\alpha}_R$ to dominate the least-squares estimator the expectation of (C.5) must be less than or equal to zero for all α, with strict inequality for some α. A sufficient condition for this is that (C.5) be negative with probability one. One further manipulation of (C.5) is required to get a condition which may be readily applied.

Substituting for C_i and letting

$$g_i = w_i \hat{\alpha}_i^2 / \sum w_j \alpha_j^2; \quad Q_i = L_i h_i^2 w_i \tag{C.6}$$

with $\sum g_i = 1$, (C.5) becomes

$$\sum Q_i t_i \{(t_i + 4) g_i / (c_i + K) - 2 / (c_i + K)\}$$

with $K = \sum \omega_j \hat{\alpha}_j^2$ and this is

$$\leq \max \{Q_i t_i (t_i + 4) / (c_i + K)\} \sum g_i - 2 \sum Q_i t_i / (c_i + K).$$

Hence a sufficient condition for $\hat{\alpha}_R$ to dominate least-squares is that

$$\max \{Q_i t_i (t_i + 4) / (c_i + K)\} < 2 \sum Q_i t_i / (c_i + K) \qquad \text{(C.7)}$$

with probability 1. Particular applications follow:

1. James–Stein estimator, James and Stein (1961):
   (a) Equal variance: $h_i = h$; $L_i = 1$; $t_i = (p - 2)$; $\omega_i = 1$; $c_i = 0$. Then $Q_i = 1$ and the condition (C.7) requires $p + 2 < 2p$ or $p > 2$, proving the dominance of the James–Stein estimator in the equal variance case.
   (b) Unequal variance modification by Efron–Morris (1972):

$$\hat{\alpha}_{iEMJS} = \{1 - (p\bar{h} - h_p)^+ / ||\alpha||^2\} \hat{\alpha}_i$$

   $L_i = 1$; $t_i = (p\bar{h} - 2h_p)^+ h_i^{-1}$; $\omega_i = 1$; $Q_i = h_i^2$; $c_i = 0$, where $(x)^+ = \max(0, x)$. Again dominance is assured in this unequal variance case provided positivity of $(p\bar{h} - 2h_p)$ holds. Thus dominance of the least-squares estimator will persist until the spectrum of $X_1^T X_1$ becomes too widely dispersed.
2. Modified Hoerl, Kennard and Baldwin rule:
   (a) Estimation. This is effected by

$$L_i = 1; \; c_i = (p - 2) h_i; \; \omega_i = 1; \; t_i = (p - 2),$$

   and the condition (C.7) for dominance is implied by

$$h_p^2 (p + 2) < 2p \overline{h^2} \qquad \text{(C.8)}$$

   where $\overline{h^2} = \sum h_j^2 / p$.
   (b) Prediction. The difference now is that $L_i = h_i^{-1}$, and the condition

$$h_p (p + 2) < 2p \bar{h}$$

   less stringent.

Appendix D
Matrix results

1. Binomial inverse theorem (see Woodbury (1950)):
   Assume matrices $A(p \times p), U(p \times q), B(q \times q)$, and $V(q \times p)$. Then, if A and B are non-singular,

$$(A + UBV)^{-1} = A^{-1} - A^{-1}UB(B + BVA^{-1}UB)^{-1}BVA^{-1}. \quad \text{(D.1)}$$

As a special case (see Plackett (1950)), take $B = I, q = 1$. A regression updating formula results if we take x_i^T to be the ith row of model matrix X $(n \times p)$, with $V = x_i$ and $U = \pm x_i$:

$$(X^T X \pm x_i x_i^T)^{-1} = (X^T X)^{-1} \mp \frac{(X^T X)^{-1} x_i x_i^T (X^T X)^{-1}}{1 \pm x_i^T (X^T X)^{-1} x_i}.$$

2. Factorization of a symmetric matrix:
   If $A(p \times p)$ is symmetric, there exists an orthogonal matrix $Q(p \times p)$ such that

$$A = Q\mathrm{Diag}(\lambda_1, \ldots, \lambda_p)Q^T \quad \text{(D.2)}$$

where $\lambda_1 \geq \lambda_2 \geq \cdots \geq \lambda_p \geq 0$. Columns of Q form an orthonormal set of eigenvectors of A. This is the spectral decomposition of A.

3. Singular value decomposition:
   This offers a more general factorization than (D.2). Let $t = \min(n, p)$, matrices $X(n \times p), U(n \times t)$, and $V(p \times t)$, then

$$X = U\mathrm{Diag}(d_1, \ldots, d_t)V^T \quad \text{(D.3)}$$

where $d_1 \geq d_2 \geq \cdots \geq d_t \geq 0$. The $r \leq t$ non-zero values d_i are the singular values of the rectangular matrix X. The squared non-zero singular values are the non-zero eigenvalues of both matrix $X^T X$ $(p \times p)$ and matrix XX^T, $(n \times n)$ both of which are of rank r. Columns of U are eigenvectors (orthonormal) of XX^T, columns of V are eigenvectors (orthonormal) of $X^T X$, corresponding to the singular values.

4. Miscellaneous:
   (a) In the notation of the singular value decomposition, (D.3), the

Moore–Penrose generalized inverse of $X(n \times p)$ of rank $r \leq \min(n, p)$ is $U \text{diag}(1/d_1, \ldots, 1/d_r, 0, \ldots, 0)V^T$.

(b) The intraclass correlation matrix A $(p \times p)$ may be written

$$A = (a - b)I + b11^T,$$

and is non-singular for $a > b$. Its eigenvalues are

$$\lambda_1 = a + (p - 1)b, \quad \lambda_2 = \cdots \lambda_p = a - b.$$

The corresponding eigenvectors are the unit vector, and any $(p - 1)$ orthogonal contrasts (orthogonal to the unit vector).

(c) Matrix P $(p \times p)$ is *idempotent* if $P^2 = P$. A symmetric idempotent matrix is called a *projection* matrix. The eigenvalues of a projection matrix are all either zero or one. The trace of the projection matrix is its rank. If P is idempotent then so is $I - P$.

Appendix E
Partial least-squares algorithm

```
plsB" <-
function(x, y, k, tol = 1e-10)
{
# S function for performing Univariate PLS based on
# Helland(1988)
# written 2nd January 1991 by M. C. Denham, University
# of Liverpool.
#
# Inputs are:
#
# x: Centred matrix (columns sum to zero) of "explanatory"
# variables
# y: vector of "responses". y must have the same length as
# number of
# rows of x and must be centred
# k: Maximum number of PLS factors to be fitted
# tol: used by lsfit to detect singularity in the matrix XW
#       (lsfit is a function in S for least-squares
# regression estimation
#       using QR decomposition)
#
#
# Output is:
#
# matrix with k columns, whose ith column is the PLS
# regression coefficient
# vector obtained by fitting i PLS factors.
#
        dimx <- dim(x)
        s <- crossprod(x, y)
        w <- s
        w <- w/sqrt(crossprod(w)[1])
        W <- w
```

```
XW <- x %*% w
b <- w %*% lsfit(XW, y, int = F)$coef
B <- b
if(k > 1) {
        for(i in 2:k) {
        w <- s - crossprod(x, (x %*% b))
        w <- w/sqrt(crossprod(w)[1])
        W <- matrix(c(W, w), nrow = dimx[2])
        XW <- matrix(c(XW, x %*% w), nrow = dimx[1])
        b <- W %*% lsfit(XW, y, int = F, tol = tol)$coef
        B <- c(B, b)
        }
}
matrix(B, ncol = k)
}
```

Bibliography

Aitchison, J. and Dunsmore, I. R. (1975). *Statistical prediction analysis*. Cambridge University Press, London.

Aitkin, M., Anderson, D., Francis, B. and Hinde, J. (1989). *Statistical modelling in GLIM*. Clarendon Press, Oxford.

Akaike, H. (1974). A new look at the statistical identification model. *IEEE Transactions on Automatic Control*, **19**, 716–723.

Allen, D. M. (1974). The relationship between variable selection and data augmentation and a method of prediction. *Technometrics*, **16**, 125–127.

Anderson, T. W. (1984). *An introduction to multivariate analysis*, 2nd edn. Wiley, New York.

Anderssen, R. S. and Bloomfield, P. (1974). A time series approach to numerical differentiation. *Technometrics*, **16**, 69–75.

Ando, A. and Kaufman, G. M. (1965). Bayesian analysis of the independent multinormal process— neither mean nor precision known. *Journal of the American Statistical Association*, **60**, 347–358.

Atkinson, A. C. (1985). *Plots, transformations and regression: an introduction to graphical methods of diagnostic regression analysis*. Clarendon Press, Oxford.

Barndorff-Nielsen, O. (1978). *Information and exponential families in statistical theory*. Wiley, Chichester.

Basu, D. (1964). Recovery of ancillary information. *Sankhya*, A, **26**, 3–16.

Bates, D. M. and Watts, D. G. (1988). *Non-linear regression analysis*. Wiley, New York.

Beale, E. M. L. (1960). Confidence regions in non-linear regression analysis (with discussion). *Journal of the Royal Statistical Society*, B, **22**, 41–88.

Becker, R. A., Chambers, J. M. and Wilks, A. R. (1988). *The New S Language: a programming environment for data analysis and graphics*. Wadsworth and Brooks/Cole, Pacific Grove, California.

Berkson, J. (1969). Estimation of a linear function for a calibration line: consideration of a recent proposal. *Technometrics*, **11**, 649–660.

Berry, D. A. (1988). Multiple comparisons, multiple tests and data dredging: a Bayesian perspective. In *Bayesian Statistics 3* (eds J. M. Bernardo, M. H. DeGroot, D. V. Lindley and A. F. M. Smith), pp. 79–94. Clarendon Press, Oxford.

Box, G. E. P. (1953). Non-normality and tests on variance. *Biometrika*, **40**, 318–335.

Brockwell, P. J. and Davis, R. A. (1987). *Time series: theory and methods.* Springer-Verlag, New York.

Brown, P. J. (1976). Remarks on some statistical methods for medical diagnosis. *Journal of the Royal Statistical Society*, A, **139**, 104–107.

Brown, P. J. (1980). Coherence and complexity in classification problems. *Scandinavian Journal of Statistics*, **7**, 95–98.

Brown, P. J. (1982). Multivariate calibration (with discussion). *Journal of the Royal Statistical Society*, B, **44**, 287–321.

Brown, P. J. and Mäkeläinen, T. (1992). Regression, sequenced measurements and coherent calibration. In *Bayesian Statistics 4* (eds J. M. Bernardo, J. Berger, A. P. Dawid and A. F. M. Smith), pp. 97–108. Clarendon Press, Oxford.

Brown, P. J. and Oman, S. D. (1990). Problematic points in nonlinear calibration. In *Statistical analysis of measurement error models and applications* (eds P. J. Brown and W. A. Fuller), pp. 139–146. American Mathematical Society, Providence, Rhode Island.

Brown, P. J. and Oman, S. D. (1991). Double points in nonlinear calibration. *Biometrika*, **78**, 33–43.

Brown, P. J. and Payne, C. D. (1975). Election night forecasting (with discussion). *Journal of the Royal Statistical Society*, A, **138**, 463–498.

Brown, P. J. and Spiegelman, C. H. (1991). Mean squared error and selection in multivariate calibration. *Statistics and Probability Letters*, **12**, 157–159.

Brown, P. J. and Sundberg, R. (1987). Confidence and conflict in multivariate calibration. *Journal of the Royal Statistical Society*, B, **49**, 46–57.

Brown, P. J. and Sundberg, R. (1989). Prediction diagnostics and updating in multivariate calibration. *Biometrika*, **76**, 349–361.

Brown, P. J. and Zidek, J. V. (1980). Adaptive multivariate ridge regression. *Annals of Statistics*, **8**, 64–74.

Brown, P. J. and Zidek, J. V. (1982). Multivariate regression shrinkage estimators with unknown covariance matrix. *Scandinavian Journal of Statistics*, **9**, 209–215.

Brown, P. J., Le, N. D. and Zidek, J. V. (1993a). Inference for a covariance matrix. In *Aspects of uncertainty: a tribute to D.V. Lindley* (eds P. R. Freeman and A. F. M. Smith). Wiley, Chichester.

Brown, P. J., Le, N. D. and Zidek, J. V. (1993b). Multivariate spatial interpolation with Kronecker covariance structures. *Canadian Journal of Statistics*, to appear.

Brown, P. J., Spiegelman, C. H. and Denham, M. C. (1991). Chemometrics and spectral frequency selection. *Philosophical Transactions of the Royal Society,* A, **337**, 311–322.

Campbell, N. A. (1980). Shrunken estimators in discriminant and canonical variables analysis. *Applied Statistics,* **29**, 5–14.

Carroll, R. J. and Spiegelman, C. H. (1986). The effect of ignoring small measurement errors in precision instrument calibration. *Journal of Quality Control,* **18**, 170–173.

Chen, Chan-Fu (1979). Bayesian inference for a normal dispersion matrix and its application to stochastic multiple regression analysis. *Journal of the Royal Statistical Society,* B, **41**, 235–248.

Clausius, R. (1850). Über die bewegende Kraft der Wärme und die Gezetze welche sich daraus für die Wärmelehre selbst ableiten lassen. *Annalen der Physik,* **79**, 368–397, 500–524.

Cleveland, W. S. (1979). Robust locally weighted regression and smoothing scatterplots. *Journal of the American Statistical Society,* **74**, 829–836.

Cook, R. D. and Weisberg, S. (1982). *Residuals and influence in regression.* Chapman and Hall, London.

Copas, J. B. (1983). Regression, prediction and shrinkage (with discussion). *Journal of the Royal Statistical Society,* B, **45**, 311–354.

Cornfield, J. (1969). The Bayesian outlook and its application. *Biometrics,* **25**, 617–657.

Cox, D. R. (1988). Some aspects of conditional and asymptotic inference: a review. *Sankhyā,* A, **50**, 314–337.

Cox, D. R. and Hinkley, D. V. (1974). *Theoretical statistics.* Chapman and Hall, London.

Cox, D. R. and Snell, E. J. (1981). *Applied statistics: principles and examples.* Chapman and Hall, London.

Cross, A. D. and Jones, R. A. (1969). *An introduction to practical infrared spectroscopy,* 3rd edn. Butterworth, London.

Davies, O. L. and Goldsmith, P. L. (1984). *Statistical methods in research and production,* 4th edn. Longmans, London.

Davis, A. W. and Hayakawa, T. (1987). Some distribution theory relating to confidence regions in multivariate calibration. *Annals of the Institute of Statistical Mathematics,* **39**, 141–152.

Dawid, A. P. (1975). On the concepts of suffiency and ancillarity in the presence of nuisance parameters. *Journal of the Royal Statistical Society,* B, **37**, 248–258.

Dawid, A. P. (1977). Spherical matrix distributions and a multivariate model. *Journal of the Royal Statistical Society,* B, **39**, 254–261.

Dawid, A. P. (1981). Some matrix-variate distribution theory: Notational considerations and a Bayesian application. *Biometrika,* **68**, 265–274.

Dawid, A. P. (1988). The infinite regress and its conjugate analysis. In *Bayesian Statistics 3* (eds J. M. Bernardo, M. H. DeGroot, D. V. Lindley and A. F. M. Smith), pp. 95–110. Clarendon Press, Oxford.

Dawid, A. P. (1993a). The island problem: coherent use of identification evidence. In *Aspects of uncertainty: a tribute to D. V. Lindley* (eds P. R. Freeman and A. F. M. Smith). Wiley, Chichester.

Dawid, A. P. (1993b). Selection paradoxes of Bayesian inference. In *Multivariate analysis and its application*. IMS Lecture-Notes Monograph Series, Hayward, California.

Dawid, A. P. and Fang, B. Q. (1992). Conjugate Bayes discrimination with infinitely many variables. *Journal of Multivariate Analysis*, **41**, 27–42.

Dawid, A. P., Stone, M. and Zidek, J. V. (1973). Marginalisation paradoxes in Bayesian and structural inference (with discussion). *Journal of the Royal Statistical Society, B*, **35**, 189–223.

de Boor, C. (1978). *A practical guide to splines*. Springer Verlag, New York.

de Finetti, B. (1974). *Theory of probability*, volume 1. Wiley, London.

Dempster, A. P. (1969). *Elements of continuous multivariate analysis*. Addison Wesley, Reading, Massachusetts.

Dempster, A. P., Laird, N. M. and Rubin, D. B. (1977). Maximum likelihood from incomplete data via the EM algorithm (with discussion). *Journal of the Royal Statistical Society, B*, **39**, 1–38.

Denby, L. and Pregibon, D. (1987). An example of the use of graphics in regression. *The American Statistician*, **41**, 33–38.

Denham, M. C. (1990). Prediction intervals in partial least squares. Technical Report. Liverpool University.

Denham, M. C. (1992). Implementing partial least squares. Technical Report. Liverpool University.

Denham, M. C. and Brown, P. J. (1993). Calibration with many variables. *Applied Statistics*, **42**, 515–528.

Dickey, J. M. (1967). Matricvariate generalisations of the multivariate t distribution and the inverted multivariate t distribution. *Annals of Mathematical Statistics*, **38**, 511–518.

Dickey, J. M., Lindley, D. V. and Press, S. J. (1985). Bayesian estimation of the dispersion matrix of a multivariate normal distribution. *Communications in Statistics: Theory and Methods*, **14**, 1019–1034.

Doob, J. L. (1953). *Stochastic processes*. Wiley, New York.

Efron, B. (1975). The efficiency of logistic regression compared to normal discrimination. *Journal of the American Statistical Association*, **70**, 892–898.

Efron, B. (1978). The geometry of exponential families. *Annals of Statistics*, **6**, 362–376.

Efron, B. and Morris, C. (1972). Empirical Bayes on vector observations: an extension of Stein's method. *Biometrika*, **59**, 335–347.

Eisenhart, C. (1939). The interpretation of certain regression methods and their use in biological and industrial research. *Annals of Mathematical Statistics*, **10**, 162–186.

Fearn, T. (1983). A misuse of ridge regression in the calibration of a near infrared reflectance instrument. *Applied Statistics*, **32**, 73–79.

Fearn, T. (1992). Shrinking calibrations. In *NIR-91 Proceedings—Making light work* (eds I. Cowe and I. Murray). NIR News, Chichester.

Fisher, R. A. (1936). The use of multiple measurements in taxonomic problems. *Annals of Eugenics*, **7**, 179–188.

Forbes, J. D. (1857). Further experiments and remarks on the measurement of heights by the boiling point of water. *Transactions of the Royal Society of Edinburgh*, **21**, 235–243.

Frank, I. R. (1987). Intermediate least squares regression method. *Chemometrics and Intelligent Laboratory Systems*, **1**, 233–242.

Friedman, J. H. (1989). Regularised discriminant analysis. *Journal of the American Statistical Association*, **84**, 165–175.

Fujikoshi, Y. and Nishii, R. (1984). On the distribution of a statistic in multivariate inverse regression analysis. *Hiroshima Mathematics Journal*, **14**, 215–225.

Fujikoshi, Y. and Nishii, R. (1986). Selection of variables in multivariate inverse regression problem. *Hiroshima Mathematics Journal*, **16**, 269–277.

Galton, F. (1889). *Natural inheritance*. Macmillan, London.

Gastwirth, J. L. (1987). The statistical precision of medical screening procedures. *Statistical Science*, **2**, 213–238.

Geisser, S. (1964). Posterior odds for multivariate normal classifications. *Journal of the Royal Statistical Society*, B, **26**, 69–76.

Gnanadesikan, R. (1977). *Methods for statistical data analysis of multivariate observations*. Wiley, New York.

Goldstein, M. and Brown, P. J. (1978). Prediction with shrinkage estimators. *Mathematische Operationsforschung und Statistik (Series Statistics)*, **9**, 3–7.

Golub, G. H., Heath, M. and Wahba, G. (1979). Generalised cross–validation as a method for choosing a good ridge parameter. *Technometrics*, **21**, 215–223.

Green, P. J. (1984). Iteratively reweighted least squares for maximum likelihood estimation, and some robust and resistant alternatives (with discussion). *Journal of the Royal Statistical Society*, B, **46**, 149–92.

Hampel, F. R., Ronchetti, P. J., Rousseeuw, P. J. and Stahel, W. A. (1986). *Robust statistics: the approach based on influence functions*. Wiley, New

York.

Harding, E. F. (1986). Modelling– the classical approach. *The Statistician*, **35**, 115–134.

Hastie, T., Tibshirani, R. and Buja, A. (1993). Flexible discriminant analysis by optimal scoring. Technical Report. AT&T Bell Laboratories, New Jersey.

Haylen, B. T., Frazer, M. I., Sutherst, J. R. and Ashby, D. (1987). The accuracy of measurement of the residual urine volume in women by urethral catheterization. Technical Report. Liverpool University.

Haylen, B. T., Frazer, M. I., Sutherst, J. R. and West, C. R. (1989). Transvaginal ultrasound in the assessment of bladder volumes in women. *British Journal of Urology*, **63**, 149–151.

Helland, I. S. (1988). On the structure of partial least squares regression. *Communications in Statistics*, **17**, 581–607.

Hoadley, B. (1970). A Bayesian look at inverse regression. *Journal of the American Statistical Association*, **65**, 356–369.

Hoerl, A. E. (1962). Application of ridge analysis to regression problems. *Chemical Engineering Progress*, **58**, 54–59.

Hoerl, A. E. and Kennard, R. W. (1970a). Ridge regression: applications to nonorthogonal problems. *Technometrics*, **12**, 69–82.

Hoerl, A. E. and Kennard, R. W. (1970b). Ridge regression: biased estimation for nonorthogonal problems. *Technometrics*, **12**, 55–67.

Hoerl, A. E., Kennard, R. W. and Baldwin, K. F. (1975). Ridge regression: some simulations. *Communications in Statistics*, **4**, 105–123.

Hoerl, A. E., Kennard, R. W. and Hoerl, R. W. (1985). Practical use of ridge regression: a challenge met. *Applied Statistics*, **34**, 114–120.

Hunter, W. G. and Lamboy, W. F. (1981). A Bayesian analysis of the linear calibration problem. *Technometrics*, **23**, 323–350.

James, W. and Stein, C. (1961). Estimation with quadratic loss. In *Proceedings of the 4th Berkeley symposium*, volume 1, pp. 361–379. University of California Press, Berkeley.

Jolliffe, I. T. (1986). *Principal component analysis*. Springer-Verlag, New York.

Jones, R. H. and Ackerson, L. M. (1990). Serial correlation in unequally spaced longitudinal data. *Biometrika*, **77**, 721–732.

Kass, R. E. and Vaidyanathan, S. K. (1992). Approximate Bayes factors and orthogonal parameters, with application to testing equality of two binomial proportions. *Journal of the Royal Statistical Society*, B, **54**, 129–144.

Kennedy, W. J. and Bancroft, T. A. (1971). Model building for prediction in regression based on repeated significance tests. *Annals of Mathematical Statistics*, **42**, 1273–1284.

Krutchkoff, R. G. (1967). Classical and inverse methods of calibration. *Technometrics*, **9**, 425–439.

Kshirsagar, A. M. (1972). *Multivariate analysis*. Marcel Dekker, New York.

Lawless, J. F. and Wang, P. (1976). A simulation study of ridge and other regression estimators. *Communications in Statistics*, **5**, 307–323.

Le, N. D. and Zidek, J. V. (1992). Interpolation with uncertain spatial covariances: a Bayesian alternative to kriging. *Journal of Multivariate Analysis*, **43**, 351–374.

Leamer, E. E. and Chamberlain, G. (1976). A Bayesian interpretation of pretesting. *Journal of the Royal Statistical Society*, B, **38**, 85–94.

Lieftinck-Koeijers, C. A. J. (1988). Multivariate calibration: a generalisation of the classical estimator. *Journal of Multivariate Analysis*, **25**, 31–44.

Lindley, D. V. (1978). The Bayesian approach (with discussion). *Scandinavian Journal of Statistics*, **5**, 1–26.

Lindley, D. V. and Smith, A. F. M. (1972). Bayes estimates for the linear model (with discussion). *Journal of the Royal Statistical Society*, B, **34**, 1–41.

Lundberg, E. and de Maré, J. (1980). Interval estimates in the spectroscopic calibration problem. *Scandinavian Journal of Statistics*, **7**, 40–42.

Lwin, T. and Maritz, J. S. (1980). A note on the problem of statistical calibration. *Journal of the Royal Statistical Society*, C, **29**, 135–141.

Mäkeläinen, T. and Brown, P. J. (1987). Priors and choice of regressors. Technical Report. University of Helsinki.

Mäkeläinen, T. and Brown, P. J. (1988). Coherent priors for ordered regressions. In *Bayesian Statistics 3* (eds J. M. Bernardo, M. H. DeGroot, D. V. Lindley and A. F. M. Smith), pp. 677–696. Clarendon Press, Oxford.

Mallows, C. L. (1973). Some comments on C_p. *Technometrics*, **15**, 661–675.

Mardia, K. V., Kent, J. T. and Bibby, J. M. (1979). *Multivariate analysis*. Academic Press, London.

Marquardt, D. W. (1963). An algorithm for least squares estimation of nonlinear parameters. *Journal of the Society for Industrial and Applied Mathematics*, **11**, 431–441.

Marquardt, D. W. (1970). Generalised inverses, ridge regression, biased linear estimation and nonlinear estimation. *Technometrics*, **12**, 591–612.

Martens, H., Jensen, S. Å. and Geladi, P. (1983). Multivariate linearity transformation for near-infrared reflectance spectrometry. In *Proceedings of the Nordic Symposium on Applied Statistics*, pp. 205–234. Stokkand Forlag, Stravanger, Norway.

Martens, H. and Næs, T. (1989). *Multivariate calibration*. Wiley, Chichester.

Matheron, G. (1971). *The theory of regionalized variables and its applications*. Centre de Geostatistique, Ecole des Mines, Paris.

McCullagh, P. and Nelder, J. A. (1988). *Generalised linear models*, 2nd edn. Chapman and Hall, London.

Minder, C. E. and Whitney, J. B. (1974). A likelihood analysis of the the linear calibration problem. *Technometrics*, **17**, 463–471.

Mosteller, F. and Tukey, J. W. (1968). Data analysis, including statistics. In *Handbook of Social Psychology* (eds G. Lindzey and E. Aronson), volume 2. Addison-Wesley, Reading, Mass.

Næs, T. and Martens, H. (1987). Testing adequacy of linear random models. *Mathematische Operationsforschung und Statistik (Series Statistics)*, **18**, 323–331.

Nishii, R. and Krishnaiah, P. R. (1988). On the moments of the classical estimates of the explanatory variables under a multivariate calibration model. *Sankhya,* A, **50**, 137–148.

Obenchain, R. L. (1977). Classical F–tests and confidence regions for ridge regression. *Technometrics*, **19**, 429–39.

Oman, S. D. (1985). An exact formula for the mean squared error of the inverse estimator in the linear calibration problem. *Journal of Statistical Planning and Inference*, **11**, 189–196.

Oman, S. D. (1988). Confidence regions in multivariate calibration. *Annals of Statistics*, **16**, 174–187.

Oman, S. D. (1991). Random calibration with many measurements: an application of Stein estimation. *Technometrics*, **33**, 187–195.

Oman, S. D. and Wax, Y. (1984). Estimating fetal age by ultrasound measurements: an example of multivariate calibration. *Biometrics*, **40**, 947–960.

Osborne, B. G., Fearn, T., Miller, A. R. and Douglas, S. (1984). Application of near infrared reflectance spectroscopy to compositional analysis of biscuits and biscuit doughs. *J. Sci. Food Agric.*, **35**, 99–105.

Payne, C., Cleave, N. and Brown, P. J. (1988). Adding new statistical techniques to standard software systems: a review. In *COMPSTAT, Copenhagen.*

Payne, C. D. and Brown, P. J. (1981). Forecasting the British election to the European Parliament. *British Journal of Political Science*, **11**, 235–245.

Plackett, R. L. (1950). Some theorems in least squares. *Biometrika*, **37**, 149–157.

Polasek, W. (1985). Hierarchical models for seasonal time series. In *Bayesian Statistics 2* (eds J. M. Bernardo, M. H. DeGroot, D. V. Lindley and A. F. M. Smith), pp. 723–732. North-Holland, Amsterdam.

Popper, K. R. (1959). *The logic of scientific discovery.* Hutchinson, London.

Press, S. J. (1982). *Applied multivariate analysis: using Bayesian and frequentist methods of inference,* 2nd edn. Krieger Publishing Company, Malabar, Florida.

Rogers, C. A. (1987). A GENSTAT macro for partial least squares analysis with cross-validation assessment of model dimensionality. *Genstat Newsletter,* **18**.

Scheffé, H. (1959). *The analysis of variance.* Wiley, New York.

Scheffé, H. (1973). A statistical theory of calibration. *Annals of Statistics,* **1**, 1–37.

Schwarz, G. (1978). Estimating the dimension of a model. *Annals of Statistics,* **6**, 461–464.

Shukla, G. K. (1972). On the problem of calibration. *Technometrics,* **14**, 547–553.

Smith, A. F. M. and Spiegelhalter, D. J. (1981). Bayesian approaches to multivariate structure. In *Interpreting Multivariate Data* (ed. V. Barnett), pp. 335–348. Wiley, Chichester.

Smith, G. (1980). An example of ridge regression difficulties. *Canadian Journal of Statistics,* **8**, 217–225.

Smith, R. L. and Corbett, M. (1987). Measuring marathon courses: an application of statistical calibration theory. *Journal of the Royal Statistical Society, Applied Statistics,* **36**, 283–295.

Spezzaferri, F. (1985). A note on multivariate calibration experiments. *Biometrics,* **41**, 267–272.

Stein, C. (1973). Estimation of the mean of a multivariate normal distribution. In *Proceedings of Prague Symposium on Asymptotic Statistics,* volume 2, pp. 345–381.

Stone, M. (1974). Cross–validatory choice and assessment of statistical predictions. *Journal of the Royal Statistical Society,* B, **36**, 111–147.

Stone, M. (1977). An asymptotic equivalence of choice of model by cross-validation and Akaike's criterion. *Journal of the Royal Statistical Society,* B, **39**, 44–47.

Stone, M. and Brooks, R. J. (1990). Continuum regression: cross-validated sequentially constructed prediction embracing ordinary least squares, partial least squares and principal components regression. *Journal of the Royal Statistical Society,* B, **52**, 237–269; Corrigendum (1992) **54**, 906-907.

Sundberg, R. (1982). When does it pay to omit variables in multivariate calibration? Technical Report. Royal Institute of Technology, Stockholm.

Sundberg, R. (1985). When is the inverse estimator MSE-superior to the standard regression estimator in multivariate controlled calibration situations? *Statistics and Probability Letters*, **3**, 75–79.

Sundberg, R. (1993). Continuum regression and ridge regression. *Journal of the Royal Statistical Society*, B, **55**, 653–659.

Sundberg, R. and Brown, P. J. (1989). Multivariate calibration with more variables than observations. *Technometrics*, **31**, 365–371.

Thisted, R. A. (1976). Ridge regression, minimax estimation, and empirical Bayes methods. Technical Report 28. Statistics department, Stanford University, California.

Thisted, R. A. (1988). *Elements of statistical computing*. Chapman and Hall, London.

Thomas, E. V. (1991). Errors-in-variables estimation in multivariate calibration. *Technometrics*, **33**, 405–413.

Tierney, L. (1990). *LISP-STAT: an object-oriented environment for statistical computing and dynamic graphics*. Wiley, New York.

Tierney, L. and Kadane, J. B. (1986). Accurate approximations for the posterior moments and marginal densities. *Journal of the American Statistical Association*, **81**, 82–86.

Tukey, J. W. (1949). One degree of freedom for non-additivity. *Biometrics*, **5**, 232–242.

Tukey, J. W. (1957). On the comparative anatomy of transformations. *Annals of Mathematical Statistics*, **28**, 602–632.

Weisberg, S. (1985). *Applied linear regression analysis*, 2nd edn. Wiley, New York.

Westlake, J. R. (1968). *A handbook of matrix inversion and solution of linear equations*. Wiley, New York.

Williams, E. J. (1959). *Regression analysis*. Wiley, New York.

Williams, E. J. (1969). Regression methods in calibration problems. In *Bulletin of the International Statistical Institute*, pp. 17–28.

Wold, S., Martens, H. and Wold, H. (1983). The multivariate calibration problem in chemistry solved by PLS. In *Matrix pencils* (eds A. Ruhe and B. Kagstrom), pp. 286–293. Lecture notes in mathematics, Springer, Heidelberg.

Wood, J. T. (1982). Estimating the age of an animal: an application of multivariate calibration. In *Proceedings of 11th International Biometrics Conference*.

Wood, J. T., Poole, W. E. and Carpenter, S. M. (1983). Validation of aging keys for eastern grey kangaroos, *macropus giganteus*. *Australian Wildlife Research*, **10**, 213–217.

Woodbury, M. (1950). Inverting modified matrices. Technical Report 42. Statistical Research Group, Princeton University.

Yaglom, A. M. (1987). *Correlation theory of stationary and related random functions*, volume 1. Springer-Verlag, Berlin.

Index

added variable plot, 17, 18
admissibility, 53
AIC, 45, 54
AIDS testing, 162
ancillarity, 88, 178
   -S, 30, 124, 178
augmented data, 66, 67, 126
autoregressive structure, 124, 125, 128, 129, 151

Bayes
   estimation, 53, 110, 117, 118
   posterior distribution, 2, 30, 61, 64, 94, 96–100, 122, 124, 126, 128, 129, 134, 144, 159, 162, 164, 166, 177, 179
   theorem, 97, 98, 161, 162, 164, 178
Beer's law, 8, 117, 131, 132, 135, 136, 153
BIC, 45
binomial inverse theorem, 102, 126, 185
bladder data, 6–8, 23, 24, 36

calibration
   controlled, 2, 22, 23, 26, 30, 31, 46, 47, 87, 92, 96, 100, 117, 135, 146
   random, 22, 23, 26, 90, 97, 100, 165
canonical
   correlations, 73, 91, 92, 98
   covariance, 72, 73
   decomposition, 58, 60, 62, 68, 79, 86, 91, 93, 115, 181
   parameter, 179
   variance, 73
Cauchy–Schwarz inequality, 71, 144
coherence, 122–125
   structural, 124
condition number, 56, 67
conjugate
   axes, 167
   gradients, 52, 77
   prior, see natural conjugate prior
contiguity, 119, 128
continuum regression, see regression,continuum
covariance kernel, 124, 133

cross-validation, 54, 55, 64, 70, 75, 81, 83, 100, 128, 129, 131, 141, 150, 152, 153, 165
   generalized, 63, 65
curvature
   intrinsic, 156

data
   bladder, 6–8, 23, 24, 36
   constructed, 56, 57
   detergent, 9, 65, 66, 68, 69, 73–76, 80, 119, 128–130, 147, 148
   Forbes, 3, 4, 14, 18
   jug and syringe, 5, 6, 18, 20
   penicillin, 4, 5, 17
   sugar, 1, 8, 10, 111, 113, 132, 136, 138, 151, 152, 154, 155
decomposition
   singular value, 58, 65, 68, 83, 185
   spectral, 52, 185
detergent data, 9, 65, 66, 68, 69, 73–76, 80, 119, 128–130, 147, 148
determinism, 121, 132, 165
diagnostics
   prediction, 82, 108
   regression, 14
diffuse prior, see vague prior distribution
discrimination, see pattern recognition
dominance, 3, 53, 63, 67, 181–183
double point, 156–159

EM algorithm, 165
errors in variables, 40, 103
estimation
   Bayes, 53, 110, 117, 118
   hyperparameter, 53, 63, 95, 100, 131, 134, 165
   maximum integrated likelihood, 64, 131, 165
   maximum likelihood, 27, 30, 32, 47, 59, 64, 70, 101, 103–105, 116, 140
   minimum unbiased risk, 64
   Stein, 3, 63, 64, 67, 131, 181, 183
evolute, 157, 159
exchangeability, 61, 77, 96, 98–100, 124, 144
exponential family